Cash Flow Planning in Agriculture

T0324160

CASH FLOW PLANNING IN AGRICULTURE

JAMES D. LIBBIN

LOWELL B. CATLETT

MICHAEL L. JONES

IOWA STATE UNIVERSITY PRESS / AMES

James D. Libbin is professor, Department of Agricultural Economics and Agricultural Business, and Extension farm management specialist at New Mexico State University.

Lowell B. Catlett is professor, Department of Agricultural Economics and Agricultural Business, and Extension marketing specialist at New Mexico State University.

Michael L. Jones is a business consultant and analyst in Indio, California.

© 1994 Iowa State University Press, Iowa 50014
All rights reserved

Authorization to photocopy items for internal or personal use, or the internal or personal use of specific clients, is granted by Iowa State University Press, provided that the base fee of $.10 per copy is paid directly to the Copyright Clearance Center, 27 Congress Street, Salem, MA 01970. For those organizations that have been granted a photocopy license by CCC, a separate system of payments has been arranged. The fee code for users of the Transactional Reporting Service is 0-8138-0642-9/94 $.10.

First edition, 1994

Library of Congress Cataloging-in-Publication Data

Libbin, James D.
    Cash flow planning in agriculture / James D. Libbin, Lowell B. Catlett, Michael L. Jones.—1st ed.
       p.   cm.
    Includes bibliographical references and index.
    ISBN 0-8138-0642-9 (alk. paper)
    1. Cash flow—Forecasting. 2. Cash management. 3. Farm income—Management. I. Catlett, Lowell B. II. Jones, Michael L. (Michael Lee), 1953– . III. Title.
HG4028.C45L497 1994
630'.68'1—dc20   93-45684

CONTENTS

Part IV. Limitations and Other Concerns, 193

PREFACE

In the Preface to *Cash Flow Forecasting* (McGraw-Hill, 1982), William Loscalzo states:

 FOR MANY PEOPLE, the transition from public accounting to industry can be unexpectedly difficult. Likewise, students making a transition to the business world will find that it is very different from the business school environment. All too often people realize that their academic backgrounds and possibly their public accounting experience have not fully prepared them to deal with the real-world problems of business. This is especially true in the area of forecasting cash flow. Those who are just beginning a career in the financial area of a corporation may very well know how to classify accounts and draft financial statements in accordance with generally accepted accounting principles (GAAP). But very often, they have no experience in how to go about preparing a forecast. In fact, they may not even know what assumptions are necessary and how to obtain them.

Similarly, few graduating seniors in agricultural economics, agricultural business, agricultural finance, or farm and ranch management are able to see the big picture, in the sense that they have not been asked to put it all together at once. Virtually all of these students have taken one or more agricultural marketing courses, a farm management course, and an agricultural finance course, and quite possibly commodity futures and farm records courses. What is often lacking is a capstone course to pull all of these subdiscipline areas together.

This book is designed for such capstone courses. It uses the cash flow forecast as the focal point and shows *how* to prepare a comprehensive, coordinated, and consistent cash flow plan. There is a great difference between knowing what a cash flow budget is (including its mechanics) and knowing *how* to begin to prepare a cash flow forecast for realistic business conditions. The Agricultural Financial Reporting and Analysis system (AFRA) is used extensively in this text as the central planning structure because it is already widely in use, it is complete and very thorough, and it is relatively easy to use.

Not only is this book useful for students in farm management and agricultural finance courses but also for farmers, ranchers, and other agriculturalists in settings like Cooperative Extension workshops, lender training courses, and educational workshops such as those offered by various professional societies.

In Part I we introduce, reintroduce, or reemphasize cash flow planning concepts and examine current planning systems. Chapter 3 contains a simple case study to show the development of a complete cash flow plan. In Part II we expand the components presented in the case study, expose the assumptions typically made when preparing cash flow plans, and investigate forecasting/planning ideas to fill the holes left by eliminating the "givens" or assumptions. The components are pulled together into a complete plan in Part III, and issues such as sensitivity to changing forecasts, risk, and monitoring are examined. We discuss the short-run cash flow plan's shortcomings, most notably the inclination of many users to equate cash flow with profitability, in the book's final chapter in Part IV.

The cash flow budget can be the central focus of all three of the major business planning problems—marketing, production, and investment—of a commercial farm or ranch. All three forms of planning are essential and are accomplished separately, but they must eventually come together to develop a comprehensive, coordinated, and consistent whole-farm or -ranch plan. Because it embodies the entire planning process, the cash flow budget is a powerful tool vital for the sophisticated business management modern farm and ranch managers require to be competitive and profitable in the 1990s and beyond.

PART I

Introduction to Cash Flow

Debt restructuring and *cash flow* are now two very common terms in U.S. production agriculture. The decade of the 1970s brought a major change to U.S. agriculture, a change which remains in force today. The financial revolution of the 1970s, due primarily to high inflation and interest rates, high fuel prices and fuel shortages, and debt financing based in large part on appreciating land values and consequent large paper equities, remains a dominant concern in farm and ranch business management. Although the forces that characterized the 1970s are not so evident in the 1990s, their aftereffects remain. The most important aftereffects are the heavy debt loads carried by a large group of U.S. farmers and ranchers and the change in emphasis by agricultural lenders from collateral-security lending to cash flow–repayment capacity lending. But not until the late 1970s or early 1980s was the impact felt to a major degree by commercial farmers and ranchers.

Heavy debt repayment commitments (both principal and interest) and low but highly variable commodity prices redirected many farmers' and ranchers' interest from profitability to cash flow. While profitability remains a major goal and is absolutely necessary for long-term growth and busi-

ness strength, cash flow is vital to short-term survival. Inadequate market planning, poor production planning, and absolutely terrible investment planning have all contributed to the severe financial crisis faced by many farmers and ranchers.

Cash flow, or flow of funds, is not a new concept by any means. Lenders, accountants, financial specialists, and farm and ranch business management specialists have long felt that the cash flow budget, including flow of funds projections, is an integral part of sophisticated business management. The primary impetus for this book is that the cash flow budget can be the central focus of all three of the major planning problems—marketing, production, and investment—of a commercial farm or ranch. All three forms of planning are essential; they must be accomplished separately at first, but eventually they must come together to develop a comprehensive, coordinated, and consistent whole-farm or -ranch plan.

The danger in putting forward the argument that the cash flow budget is the central focus of farm and ranch planning is the implication that this statement is more important than the other financial statements (balance sheet, income statement, statement of owner equity, and statement of cash flows)[1] or that this planning process is more important than market planning or production planning. We will attempt to avoid this danger since the cash flow budget is but one vital piece of information. We will argue instead that the cash flow budget can be a very powerful tool since it actually embodies the *entire* planning process. Adequate market and production planning must be done first, well before a comprehensive cash flow plan can be developed.

The cash flow plan not only embodies all of the primary types of business planning, it also summarizes the one hard-and-fast, inviolable rule of business management: If a business cannot generate the cash to meets its commitments, *it cannot survive*. An unprofitable business can survive, a technically bankrupt farmer or rancher can remain in business, and a poor marketer can survive *if* the business has a source of owned or borrowed cash with which to continue. However, a negative bank balance with no further means to gather cash is the single most important reason why businesses dissolve.

NOTE

1. The difference between a cash flow budget (or statement) and a statement of cash flows is probably hazy to all who have not formally studied the construction of the relatively new Statement of Cash Flows as developed in Financial Accounting Standards Board Statement No. 95. The differences between the two statements are highlighted in Chapter 2.

CHAPTER 1

The Importance of Cash Flow

Financial control, forward planning, and risk management are three of the most difficult and troublesome tasks faced by modern commercial farmers and ranchers for several reasons. Financial planning, record keeping, and financial control are recognized as important by most farmers and ranchers, but very few of these individuals *want* to maintain records or to spend a great amount of their time on financial matters. These challenges were not the primary reasons why farmers and ranchers began farming or ranching in the first place—if they had preferred accounting and finance, they would have probably chosen accounting or finance professions rather than farming or ranching. Farmers and ranchers typically prefer working with their hands, producing crops and livestock, or repairing machinery, to pencil pushing. Furthermore, peering into the future is a frustrating task given all of the uncertainties that we know we will have to face; uncertainties concerning weather conditions that will affect crop yields and livestock performance, uncertainties about where market prices are headed, and uncertainties about the costs, timing, and availability of major inputs are just a few examples. It is too easy, however, to allow yourself to get so frustrated with the effects of these uncertainties that the planning process is abandoned before it begins or before it is given adequate contemplation.

Few of us have taken the time to lay out on paper what our short-term or long-term goals are for our farms, ranches, careers, personal lives, or families. But yet, each of us lays out some set of goals for the future, even if it is done only subconsciously or just for one day. Have you never contemplated what you want to accomplish during the day while doing morning chores, eating breakfast, or driving to school or work? Have you never contemplated which route to take on a trip and how long it will take or how much money you will need? If you have ever thought about any of these things, then you are well on the road to being a planner.

Discussions about business planning often use a long automobile trip as an analogy. Did you ever take a long trip without knowing where you want to go? Did you just get in the car and take off? Did you think about filling the fuel tank, packing clothes and other necessities, and getting cash or traveler's checks before you leave home? Are the spare tire, jack, and lug wrench in the car and operable? Do you know in which direction to head? Did you study a road map to determine which highways will get you to your destination? Have you thought about how long it should take you to get there?

Even the novice business planner knows that this analogy is so simple that it is not relevant. Why, planning a business is so much more complex and difficult that this analogy insults our intelligence, doesn't it? Trip takers never have to contend with setting long-term goals, contemplating what to produce, arranging for physical and financial inputs, adjusting to risks, setting short-term goals, determining how to produce, or thinking about the problems that might face the business along the way.

If you are willing to accept that there are some major differences in scale (a business involves many more decisions and a longer time frame) and consequences (an extra day on a trip may not be so bad, but a bad decision might lead to severe financial loss to a business), and lay out the questions posed about the trip and the problems faced by a business as we have done in Figure 1.1, you may see that the analogy is not so bad after all. Trip takers and business people both need to know where they are going or hope to go; in other words they have to set goals. Few farmers drop the chisel into the ground without having some idea of what they are going to plant just as few travelers get in the car and go without knowing where they are going. Every trip and every farm seem to need money, fuel, and other items such as feed, fertilizer, clothes, or toys for the kids. Farmers and ranchers face many uncertainties, and cars on trips seem to attract nails on the road. Drivers need to know compass directions and businesses need to have short-run goals. Drivers need to know how to get there and so do farmers and ranchers. Finally, both travelers and business owners need to know what they are doing, how far they can go, and what they can truly accomplish.

WHAT IS CASH FLOW?

A cash flow statement, budget, or plan[1], stated about as simply as we can make it, is a moving picture of the movement of cash into and out of a business. The cash flow budget puts in writing, on one piece of paper, how much cash you have to begin

Trip Questions	Business Problems
1. Where to go?	1. Set long-term goals.
2. Get in and take off?	2. Determine what to produce.
3. Fuel, clothes and money?	3. Arrange for inputs.
4. Spare time and tools?	4. Face and plan for uncertainties.
5. Which direction?	5. Set short-term goals.
6. Road map?	6. Determine how to produce.
7. How long?	7. How much progress can be accomplished?

Figure 1.1. Taking an automobile trip versus running a business.

with, where cash inflows will come from, where cash outflows will go, and how much cash will be left over at the end. The cash flow budget clearly addresses one of the most serious financial problems faced today—cash flow control. The cash flow budget is also a primary element in the credit application process because it directly shows how and when borrowed capital (especially short-term operating loans) will be repaid. In both cases, the farmer or rancher develops a cash flow budget because he recognizes the cyclical flows of cash into and out of an agricultural business. Typically, cash needs are greater than the amount of cash available during productive stages of animal and plant growth. Moreover, sales tend to be lumpy; they do not flow in evenly throughout the year but arrive in bunches and come *after* the production takes place. As a result of both cash inflow and outflow patterns, periods of cash deficits exist that cause many farmers and ranchers either to draw down savings accounts or to borrow to cover the deficits.

Perhaps the most important feature of the cash flow budget is that it can be used as the culmination of the annual farm or ranch marketing/financial planning process. If the cash flow budget is prepared with a great deal of thoughtful planning, it can wrap together the producer's marketing plans, optimal enterprise selection and input needs, family living needs, feed requirements, inventory sources and uses, capital expansion plans, and debt repayment capacity.

There is no doubt that the cash flow budget is a very important and useful financial management tool. However, it requires a lot of careful work (see Figure 1.2) and clearly defined plans. Conceptually, the cash flow budget is easy to understand. Practically, it is difficult to draw up, not because it involves many difficult or abstract concepts, but because it does require extensive planning.

The cash flow budget addresses a primary fact of life—that negative bank balances are not very popular with bankers. In fact, the major objective of the cash flow budget is to show the sources of cash generated by the business and the cash needs or requirements of the business. In periods when cash needs exceed cash sources, money must be borrowed. When cash sources exceed needs, the excess can be carried over to cover future needs or deficits, or it can be used to purchase additional supplies or capital assets, increase savings, or retire existing debts. Regardless of whether there is an excess of cash sources over requirements and regardless of which purpose the excess is committed to, the summation of cash sources must be equal to the summation of cash uses. In other words, every dollar flowing into the business must be allocated to some use, and, regardless of whether money is earned or borrowed, there must be enough of it available to cover all cash needs.

PROJECTED VERSUS ACTUAL CASH FLOW

The cash flow budget is somewhat of an oddity with respect to the other financial statements. Very seldom is a cash flow budget prepared on an actual basis, that is, to show what has already happened. Almost all are prepared on a projected basis—they are developed to show how the business is *expected* to perform in the future. Most balance sheets are prepared on an actual basis to show the position of the business on a particular day. Similarly, most income statements (an accrual-based determination of profitability) and statements of cash flows are prepared on an actual basis to show

19X0 CASH FLOW BUDGET

DESCRIPTION	TOTAL	JAN	FEB	MAR	APRIL	MAY	JUNE	JULY	AUG	SEPT	OCT	NOV	DEC
BEG. CASH BALANCE	800	800	1000	1000	1000	1000	1000	1000	1000	1000	1000	1000	1000
OPERATING RECEIPTS:													
CROPS AND FEED	213605	0	13500	0	0	0	0	47500	0	42500	39825	70280	0
LIVESTOCK & POULTRY	34930	15834	19096	0	0	0	0	0	0	0	0	0	0
CUSTOM WORK	7000	0	0	0	0	0	0	3000	0	4000	0	0	0
CAPITAL RECEIPTS:													
BREEDING STOCK	850	850	0	0	0	0	0	0	0	0	0	0	0
MACHINERY & EQUIP.	0	0	0	0	0	0	0	0	0	0	0	0	0
NON-FARM INCOME:													
INTEREST	496	0	0	0	0	0	0	0	0	0	0	0	496
OFF-FARM WAGES	9000	0	1500	0	1500	0	1500	0	1500	0	1500	0	1500
TOTAL CASH AVAILABLE	266681	17484	35096	1000	2500	1000	2500	51500	2500	47500	42325	71280	2500
OPERATING EXPENSES:													
LABOR HIRED	18000	1500	1500	1500	1500	1500	1500	1500	1500	1500	1500	1500	1500
REPAIRS-MACH & EQUIP.	8000	0	0	1000	2500	0	0	0	1500	1000	2000	0	0
REPAIRS-BUILD/IMPROV.	2500	0	0	0	0	1200	0	0	0	0	0	1300	0
RENTS & LEASES	62400	0	0	31200	0	0	0	0	0	31200	0	0	0
SEED	11860	0	0	9540	1920	0	0	0	0	400	0	0	0
FERTILIZER & LIME	40300	0	0	32260	3200	0	0	0	0	0	0	4840	0
CHEMICALS	11730	0	0	0	10557	1173	0	0	0	0	0	0	0
LIVESTOCK EXPENSE	5000	1250	0	0	1250	0	0	1250	0	0	1250	0	0
GAS, FUEL, OIL	12000	500	0	1000	2000	2000	3000	0	0	500	1000	2000	0
STORAGE/CUSTOM DRY	4500	0	0	0	0	0	0	0	0	0	3500	1000	0
TAXES (REAL EST, PP)	6000	0	0	0	0	0	0	0	0	6000	0	0	0
INSURANCE(PROP,LIAB)	1400	0	0	0	0	0	0	0	0	0	0	0	1400
UTILITIES(ELECT/GAS)	6000	500	500	500	500	500	500	500	500	500	500	500	500
AUTO (FARM SHARE)	1800	150	150	150	150	150	150	150	150	150	150	150	150
TOTAL CASH OPER EXPS	191490	3900	2150	77150	23577	6523	5150	3400	3650	41250	9900	11290	3550
STOCK & FEED PURCH:													
FEED PURCHASED	18000	0	0	18000	0	0	0	0	0	0	0	0	0
CAPITAL EXPENDITURES													
MACHINERY & EQUIP	35000	0	0	0	0	0	0	0	0	0	0	35000	0
OTHER EXPENDITURES:													
GROSS FAMILY LIV W/D	21600	1800	1800	1800	1800	1800	1800	1800	1800	1800	1800	1800	1800
INCOME TAX & SOC SEC	10659	0	0	10659	0	0	0	0	0	0	0	0	0
LOAN PAYMENTS - PRIN	27861	0	0	0	0	0	0	3800	19883	0	0	4178	0
LOAN PAYMENTS - INT	15493	0	0	0	0	0	0	8185	5164	0	0	1894	250
TOTAL CASH REQUIRED	320103	5700	3950	107609	25377	8323	6950	17185	30497	43050	11700	54162	5600
CASH AVAIL - CASH REQ	-53422	11784	31146	-106609	-22877	-7323	-4450	34315	-27997	4450	30625	17118	-3100
INFLOWS FROM SAVINGS	67390	1000	0	0	0	0	0	0	28997	0	0	11939	25454
CASH POS BEFORE BORR	13968	12784	31146	-106609	-22877	-7323	-4450	34315	1000	4450	30625	29057	22354
MONEY TO BE BORROWED													
-OPERATING LOANS	166326	10894	10173	107609	23877	8323	5450	0	0	0	0	0	0
-INT & L/T LOANS	25000	0	0	0	0	0	0	0	0	0	0	25000	0
OP LOAN PAY - PRIN	118565	18000	39978	0	0	0	0	0	0	0	0	40000	20587
-INTEREST	19339	4678	341	0	0	0	0	0	0	0	0	13057	1263
OUTFLOWS TO SAVINGS	66390	0	0	0	0	0	0	33315	0	3450	29625	0	0
ENDING CASH BALANCE	1000	1000	1000	1000	1000	1000	1000	1000	1000	1000	1000	1000	1000
LOAN BALANCES:													
CURRENT YR'S OP LOAN	105739	10894	21067	128676	152553	160876	166326	166326	166326	166326	166326	126326	105739
PREV YR'S OPER LOANS	0	39978	0	0	0	0	0	0	0	0	0	0	0
INT & LONG TERM LOAN	190169	193030	193030	193030	193030	193030	193030	189230	169347	169347	169347	190169	190169
TOTAL LOANS	295908	243902	214097	321706	345583	353906	359356	355556	335673	335673	335673	316495	295908

Figure 1.2. Example of a completed farm cash flow budget.

how the business performed over the past year (or some other period).

All four statements, however, can be extremely valuable management tools if they are projected into the future. First of all, recognize that all four statements describe the same business and thus should be chapters of the same story, telling that story from a different viewpoint. Each of these viewpoints is necessary to adequately discover business opportunities and problems. If we argue that the cash flow budget represents or embodies the entire business planning process, then a projected balance sheet, income statement, and statement of cash flows could be said to show the consequences of that plan.

By definition, a cash flow plan or budget must be a future projection (seldom will we attempt to project the past). An actual cash flow statement would show how and when money was received or spent in the past and would show historical bank balances; its information, other than the monthly components, is almost totally embodied in the statement of cash flows. Last year's cash flow statement (or a budget prepared at the beginning of last year) can be a very effective guide for beginning this year's cash flow plan. Most businesses tend to settle into a production, purchasing, and marketing pattern that seems to fit the geographic area, limitations on crop and livestock production, funds availability, and common practices. Each one of these activities or assumptions should be challenged when preparing a projection, but past activities can be used as a starting point or guide to future activities. Preparation of an actual cash flow statement will require very little additional effort if the cash flow budget developed at the beginning of the year is monitored throughout the year, as will be discussed in Chapter 11.

Usually, cash flow budgets are prepared as plans or guidelines for each month of the next calendar year. Cash flow plans can be prepared farther in advance than one year, but the farther into the future the plan is projected, the more tentative become the results because of the increasing likelihood of unanticipated events. Consequently, cash flow plans projected more than one year in advance typically are used for major capital expansion plans and will usually relate only to the purchase of that capital asset, along with the additional cash sources and uses generated by the expansion. This chapter will be confined to a typical, short-term (12-month) cash flow budget.

CASH FLOW AND PROFITABILITY

Despite the recent emphasis by agricultural lenders on cash flow planning and debt repayment capacity, the cash flow budget represents only one part of the financial management picture—a very important part of that picture, but nevertheless only one part. Much of the farm or ranch records system must be maintained on the accrual accounting system or converted to the accrual system through inventory adjustments. The primary purpose for this process is that the cash accounting system cannot adequately reflect how profitable a business is because the timing of purchases and sales should not affect the underlying use of inputs or the production of income.

Within the cash flow budget, only cash inflows and outflows count. Consequently, the cash flow budget says nothing about the profitability of the business. Profitability information is the task of the income statement, the cash flow budget describes cash position, and the balance sheet shows overall financial position and strength. All of the statements are necessary as each has a different mission or emphasis.

The cash flow budget holds a unique position within the set of financial statements because it forces recognition of the manipulations necessary to maintain a positive bank balance. The world moves on the basis of cash. Bankers and other suppliers no longer accept payment in commodities. A net loss in one year will not necessarily force a business to fail, but a cash deficit or the inability to raise enough cash to meet obligations will cause failure. And yet a positive cash flow is not sufficient by itself.

Eventually, the business must be profitable for it to continue.

STRUCTURE OF THE CASH FLOW BUDGET

A cash flow budget typically can be broken into five major parts.

1. Beginning cash balance (the amount of cash in currency or checking accounts ready for use)
2. Receipts (the money inflows generated by sales, off-farm employment, and other income)
3. Expenditures (the money outflows needed to operate the business)
4. Adjustments (additional cash inflows to cover deficits or cash control maneuvers)
5. Ending cash balance

A typical example of a general cash flow budget is shown in Figure 1.3. The beginning cash balance is added to receipts to calculate the total cash available for the period. From the cash available, expenditures are subtracted. The balance will be either positive or negative. If negative, a cash deficit exists and somehow must be covered. Typically, cash will either be withdrawn from savings or it will be borrowed to cover the deficit and provide some operating reserves for the next period. (If additional income is generated through early sales to cover the projected deficit, those sales would be reported in the receipts section.) If cash available less cash required results in a cash surplus, or positive balance, then the manager must decide whether he or she wants to place the excess into savings (an interest-bearing account), repay outstanding operating loans (or possibly prepay longer-term loans), or carry the positive balance into the next period. After these savings/borrowing/repayment adjustments, the amount left over is the ending cash balance. That ending cash balance will then become the beginning cash balance for the next period (shown as A in Figure 1.3).

Individual farmers and ranchers will want to develop their own cash balance strategy for determining how much cash is to be carried over from one period to the next. A certain minimum amount will usually be determined to provide a cushion at the beginning of the next period. Within a period, as well as between periods, cash flow inequities can occur. For example, Period 2 sales may not be slated until the middle of the period, but some of the bills are due at the beginning of the period.

In addition to this type of within-period cash flow problem, the cash balance strategy should reflect any notice required before savings can be withdrawn, whether a cash excess or deficit is projected for the next few periods, and the type of operating loan structure under which short-term borrowings are made. If short-term money is borrowed under a revolving line of credit, repayments should be made as quickly as possible. A revolving line of credit specifies the maximum amount of borrowed money that can be outstanding at any one time. Money can be borrowed when needed, repaid, and at a later date borrowed again, as long as the maximum outstanding limit is not exceeded. Since interest charges are always assessed against the outstanding balance, it will be to the borrower's advantage to keep that outstanding balance as low as possible. If money is not borrowed under a revolving line of credit (either through

	Period 1	Period 2	...
Beginning Cash Balance		A	
+ Receipts			
= Total Cash Available			
− Expenditures			
= Cash Available Less Cash Required			
+ Inflows from Savings			
+ Cash Borrowed			
− Short-term Loan Repayments			
− Outflows to Savings Accounts			
= Ending Cash Balance	A		

Figure 1.3. General structure of a cash flow budget.

a nonrevolving line or a fixed amount loan), the funds cannot be reborrowed. Consequently, such borrowers will want to be more conservative in their repayment plans to protect against future contingencies.

AGRICULTURAL FINANCIAL REPORTING AND ANALYSIS SYSTEM

A system of coordinated financial statements called the Agricultural Financial Reporting and Analysis system (AFRA) developed by Arnold W. Oltmans, Danny A. Klinefelter, and Thomas L. Frey will be used exclusively in this text.[2] The use of AFRA is not intended to promote a specific product, but rather, AFRA is widely recognized as a complete, thorough, and coordinated set of financial statements. Surveys of students learning AFRA for the first time indicate that AFRA is difficult to learn, but once learned, the package offers great insight into the inner workings of a business.

The AFRA package includes a balance sheet, an income statement, a cash flow budget, a statement of cash flows, a statement of owner equity, and an analysis section—the 16 measurements and ratios recommended by the Farm Financial Standards task force.[3] Although the cash flow budget is the focus of this text, all four items will be discussed in the next chapter.

COMPUTERIZATION

The calculations that must be performed to complete a cash flow budget are simple conceptually: addition and subtraction will get the job done. The problem is that there are hundreds or even thousands of cumulative calculations to perform. The simple structure of the statement and the types and amount of calculations to be performed readily suggest that a computer could be put to good use. Several excellent computer packages, usually spreadsheet programs, are available.[4] The computer can be an extremely valuable tool, especially in completing the mechanics of the cash flow budget and in considering alternatives.

SUMMARY

Cash flow planning and control are vital areas of farm and ranch business management. These areas really have always been important for farmers and ranchers but only recently has the importance of cash flow as the lifeblood of a business become so clear.

Cash flow budgets can be prepared in two different manners: projected or actual. A projected budget should be thought of as a plan for the near future. This plan must be monitored throughout the time period; that is, it should be checked to see how closely actual conditions compare with projections. The purpose of this comparison is not to test the skills of the planner, but rather to decide whether further adjustments or plans are needed to respond to changing conditions. An actual cash flow statement is easy to prepare, it completes or closes the cash transactions portion of the record books, and it provides an excellent starting point for preparing the next cash flow budget.

The underlying structure of the cash flow budget can be summarized in one simple equation: sources of funds = uses of funds. Sources of funds include beginning balances, cash receipts from business or nonbusiness sources, and loan receipts. Uses of funds include expenditures, family withdrawals, investments, and cash carryovers. Adjustments, such as savings account balance and operating loan changes, even out the flow of funds.

Cash flow planning is a long, difficult, and tedious process if it is done seriously and must reflect all of the near future plans of the business. Many estimates and outright guesses must be made that may seem to negate the value of the exercise. No prognosticator, however, can see into the future with perfect accuracy. A good plan, or road map into the future, will help to define both goals and paths to meet those goals.

A careful understanding of the difference between the income statement and the cash flow budget is absolutely necessary. The income statement is an accrual statement that portrays profitability. The cash flow budget is totally cash oriented and depicts liquidity or the ability to meet cash obligations as they come due.

NOTES

1. Throughout this text we will use the words *budget* and *plan* interchangeably; a budget or a plan is a forward look or projection.

2. A.W. Oltmans, D.A. Klinefelter, and T.L. Frey, *Agricultural Financial Reporting and Analysis,* Century Communications, Inc., Niles, Ill. 60714, 1992 (phone 708-647-1200).

3. Additional discussion of the roles and preparation of AFRA (formerly CFS) statements can be found in J.D. Libbin and L.B. Catlett, *Farm and Ranch Financial Records,* Macmillan, 1987; A.G. Nelson and T.L. Frey, *You and Your Cash Flow,* Agri-Business Publications Division, Century Communications, 1983; and J.B. Penson, Jr., and C.J. Nixon, *Understanding Financial Statements in Agriculture,* Agri-Information Corporation, 1983.

4. See J.T. McGuckin, *NMSU Cash Flow Worksheet,* DS-2, Cooperative Extension Service, New Mexico State University, June 1984; FBS Systems, Documentation for Agricultural Financial Reporting and Analysis software, FBS Systems, Inc., Aledo, Ill. 61231, 1994.

RECOMMENDED READINGS

David A. Lins, "Cash Flow Planning," *Money Sense*, John Deere Corporation, Moline, Ill., Fall 1984.

Gerald G. Geisler, "Cash Flow Aids Farm Planning," *Southwest Farm Press*, Clarksdale, Miss., March 21, 1985, p.15.

James D. Libbin and Lowell B. Catlett, "Does Business Management Make a Difference?" *AgriCents*, Cooperative Extension Service, New Mexico State University, Las Cruces, N.Mex., Summer 1987.

A. Gene Nelson and Thomas L. Frey, *You and Your Cash Flow*, Century Communications, Inc., Skokie, Ill., 1983.

FBS Systems, Documentation for Agricultural Financial Reporting and Analysis software, FBS Systems Inc., Aledo, Ill., 1994.

Arnold W. Oltmans, Danny A. Klinefelter, and Thomas L. Frey, *Agricultural Financial Reporting and Analysis*, Century Communications Inc., Niles, Ill. 1992.

James D. Libbin and Lowell B. Catlett, *Farm and Ranch Financial Records*, Macmillan, New York, N.Y., 1987.

J. Thomas McGuckin, *NMSU Cash Flow Worksheet*, DS-2, Cooperative Extension Service, New Mexico State University, Las Cruces, June 1984.

W.M. Greenfield and Dennis P. Curtin, *Cash Flow Management with Framework*, Curtin & London, Inc., Marblehead, MA/Prentice-Hall, Inc., Englewood Cliffs, N.J., 1986.

Charles W. Kyd, "Getting the Cash Out of Cash Flow," *Inc.*, July 1987, pp. 87-88.

J.W. Prevatt, *Cash Flow Analysis: A Farm Management Technique*, Circular 488, Cooperative Extension Service, University of Florida, Gainsville, Fla., 1981.

CHAPTER 2

Relationships between the Financial Statements

In Chapter 1 we argued that each one of the four financial statements was built upon different assumptions and that each serves a very different purpose. Although that argument is very true, it should also be evident that these financial statements are not developed in a vacuum, each relates to the same business, and in fact the same numbers from the records system are often used in two separate statements and sometimes in all four.

The primary purpose of this chapter is to explore the relationships among the four most important financial statements: the balance sheet, the income statement, the statement of cash flows, and the cash flow budget. We will also explore the role of the statement of owner's equity in reconciling accrual income to change in net worth. Once established, the use of these relationships will be extended to the development of projected or pro forma statements. But first we should review the basic assumptions of each of the statements.

The balance sheet is the classic representation of the fundamental accounting equation at one point in time. It is a stock statement that describes the financial position of the business and/or family unit on a particular day. Because of the heavy reliance in agriculture on fixed, expensive, long-term assets such as land, buildings, and machinery, assets whose values have been greatly influenced by inflation, a double-column (market value and modified cost) balance sheet is necessary to maintain relationships with the income statement and to recognize the different needs of different users of the statement.[1] Often the balance sheet is developed on a consolidated family-business basis, usually because the owner will prepare only one balance sheet, because the owner often operates the farm or ranch as an integral part of family life, and because the assets and liabilities of a typical farm or ranch may often be difficult to split between business and family.

In contrast to the balance sheet, the income statement is a flow statement, that is, it is prepared to represent a period of time rather than a point in time. A farm or ranch income statement is usually prepared to cover a period of one year because the normal production cycle in agriculture makes more frequent statements difficult to estimate. However, for more continuous production patterns such as dairy or year-round–confinement hog or poultry production, more frequent income statements become easier if they are needed. The income statement must be developed on an accrual basis to be useful. A cash-based receipts-expenditures records system can be converted to an accrual income statement through recognition of changes in inventories, accounts payable and receivable, and prepaid and accrued expenses.

Like the income statement, the statement of cash flows is also a flow statement that represents a period of time. In contrast to the income statement, however, the statement of cash flows is relatively new (it was adopted by the Financial Accounting Standards Board in 1987[2]), not widely used or well understood (mainly because it is still relatively new), and, most importantly, strictly cash based. The primary purpose of the statement of cash flows is two-fold: to reconcile accrual-based net income to change in cash position (and thus further emphasize the difference between cash and accrual) and to help users of financial information understand seemingly conflicting results such as a company taking a loss in one year and yet purchasing a new subsidiary. The actual statement of cash flows can be an extremely valuable tool, serving as a summary of the year's transactions. This summary will provide a great share of the detail necessary for income tax reporting, income statement preparation, and debt reconciliation.

The statement of owner equity, the unincorporated business analog to the statement of retained earnings, reconciles the level of net worth at the end of the year to that at the beginning of the year by adjusting beginning net worth for net income, withdrawals, and other transfers.

The cash flow budget is also strictly cash based. It represents a monthly summary of cash receipts and cash expenditures as well as the steps taken, such as new borrowing and repayment of operating loans or flows of cash reserves between the savings and checking accounts, to adjust for differences between cash inflows and outflows. Like the income statement and statement of cash flows, the cash flow budget is a flow statement; it is prepared to cover a period of time. Once again, that period is usually one year, although it can be projected further into the future. A cash flow budget is most useful if projected for one year on a monthly basis, used as a budget or plan for cash flows for the next production year, and then monitored during the year.

THE BALANCE SHEET AND THE INCOME STATEMENT

To best understand the relationship between the balance sheet and the income statement, recall the fundamental accounting equation:

$$Assets = Liabilities + Net\ Worth$$

Each type of account has a direct effect on the fundamental equation. A pictorial representation of the relationships between types of accounts is shown in Figure 2.1.

The formalized, formatted summary of the activity recorded in the individual income and expense accounts is the income statement. Although that step may be clear, the next step of the relationship between the income statement and the balance

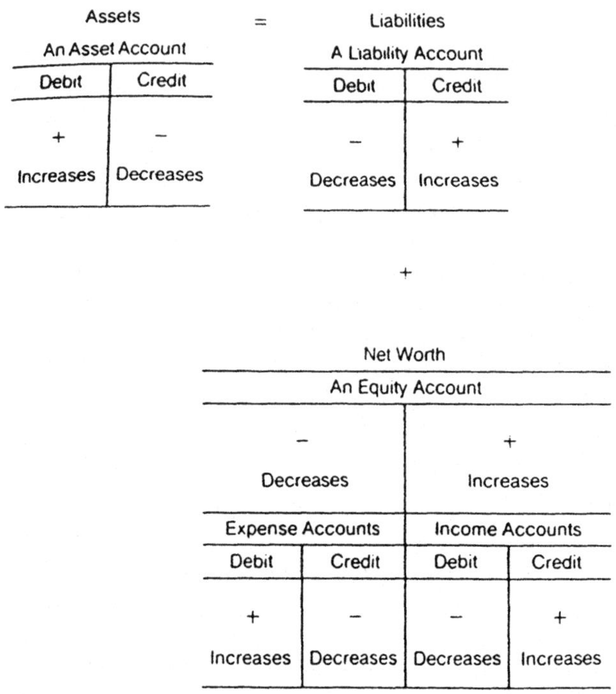

Figure 2.1. **Relationship of types of accounts to the fundamental accounting equation.**

sheet may not be so evident. By the position of the income and expense accounts in Figure 2.1, it is easy to see that changes in income and expenses directly translate into changes in net worth. In fact, all of the changes in net worth occur as a result of activity in the income and expense accounts with the exception of money withdrawn from the business or money added from an outside source.

To confirm that this statement is true, ask yourself the question: How can net worth be increased? The most often heard reactions are to buy new assets or to pay off existing debts. But if we analyze these two reactions, both can be shown to have no effect on net worth.

If new assets are purchased, say 100 acres of land are added to an existing farm, the land account increased by the value of the 100 acres. Temporarily the asset side of the fundamental accounting equation increases. But what paid for that land? If cash was paid, then the cash account decreased by the same amount as the land account increased. Consequently total assets net out the same as they were before the land purchase and more importantly, net worth has not changed. If the land purchase was financed, then the mortgages payable account increases by the same amount as the land account increased. Consequently the fundamental equation remains in balance with no effect on net worth.

Consider what happens when a debt is paid off. The note payable account will decrease by the amount of the principal balance due. Although liabilities decrease, the cash account will decrease by the same amount and the equation will balance without effect on net worth. These two reactions, or cases, are set out in Figure 2.2.

Case 1: Buy assets (land)

a. Cash purchase of $200,000

Land	Cash	NW
$200.000|	|$200,000	0 0

b. Financed purchase

Land	Mortgages Payable	NW
$200,000|	|$200.000	0|0

Case 2: Repay Existing Debt ($100,000)

Cash	Mortgages Payable	NW
|$100.000	$100.000|	0 0

Case 3: Net Farm Income is $20,000

Crop Expenses	Crop Income	=	NW
|$80,000	$100,000|		|$20,000

Cash
$20.000|

Case 4:   Gift of $10,000 to the Business

Cash	Capital-NW
$10.000|	|$10.000

Figure 2.2. Effect of transactions on net worth.

In contrast to these two cases, which have no effect on net worth, consider a year in which crop income exceeded expenses by $20,000 (Net Farm Income on the income statement is $20,000). The crop income account increases and the cash account increases (if the crop is sold). Income is wrapped into net worth through the chart of Figure 2.1. The net result is that assets increase by $20,000, and net worth increases by $20,000 (Case 3 in Figure 2.2). No other activity within the business can increase net worth!

There are other ways to affect net worth, but each of these ways implies transactions between the business and an outside agent. Net worth can be increased through inheritances or gifts or through outside investment dollars flowing into the business. Net worth will be decreased by gifts to others or through owner withdrawals for family purposes. A gift of $10,000 to the business is shown as Case 4 in Figure 2.2.

STATEMENT OF OWNER EQUITY

The statement of owner equity (Figure 2.3) presents a convenient location to summarize the relationship between the income statement and change in net worth. Mathematically, that relationship is:

STATEMENT OF OWNER EQUITY

SOE

For 12-month period Ending _____ , 19____

Name _____

Address _____ Phone _____

	Cost	Market Value
TOTAL OWNER EQUITY, Beginning of Period (beginning Balance Sheet, line 58) .... (1)	$ _____	$ _____

Change in Contributed Capital and Retained Earnings:

Net Income (loss) after taxes for the period (Income Statement, line 33) .... $ ± _____ (2)

Withdrawals of net income and retained earnings (cash or property) during the period:

    Withdrawals for family living expenses .... $_____ (3a)

    Withdrawals for investments into personal assets .... _____ − _____ (3)

Additions of capital (cash or property) to the business during the period:

    Gifts & inheritances received; additions to paid-in capital .... $_____ (4a)

    Investments of personal assets into the business .... _____ + _____ (4)

Distributions of capital, dividends, or gifts made (cash or property) during the period ...... − _____ (5)

    TOTAL CHANGE IN CONTRIBUTED CAPITAL AND RETAINED EARNINGS (add lines 2 through 5) .... (6) ± _____ ± _____

Change in Personal Net Worth:

Personal net worth, end of period (ending Balance Sheet, line 56) .... $ + _____ (7)

Personal net worth, beginning of period (beginning Balance Sheet, line 56) .... − _____ (8)

    TOTAL CHANGE IN PERSONAL NET WORTH (line 7 minus line 8) .... (9) /////////// ± _____

Change in Valuation Equity:

Valuation equity, end of period (ending Balance Sheet, line 57) .... $ + _____ (10)

Valuation equity, beginning of period (beginning Balance Sheet, line 57) .... − _____ (11)

    TOTAL CHANGE IN VALUATION EQUITY (line 10 minus line 11) .... (12) /////////// ± _____

TOTAL OWNER EQUITY, End of Period (as calculated) (add lines 1, 6, 9, and 12) .... (13) $ _____ $ _____

TOTAL OWNER EQUITY, End of Period (as reported) (ending Balance Sheet, line 58) .... (14) $ _____ $ _____

FOOTNOTES TO STATEMENT OF OWNER EQUITY

Comparison of calculated Owner Equity to reported Owner Equity, End of Period

	Cost	Market Value
Difference between Owner Equity (as calculated) and Owner Equity (as reported) End of Period: (line 13 minus line 14)	$ _____	$ _____

Explanation of difference (if applicable):

(a) Over (under) reported withdrawals of net income:

(b) Over (under) reported additions and distributions of capital into the business:

(c) Unaccounted errors in reported income or expenses:

(d) Inter-year adjustments to asset and liability values:

(e) Other:

CHANGE IN TOTAL OWNER EQUITY (as calculated) from beginning to end of period (line 1 minus line 13) .... $ _____ $ _____

CHANGE IN TOTAL OWNER EQUITY (as reported) from beginning to end of period (line 1 minus line 14) .... _____ _____

Figure 2.3. AFRA statement of owner equity.

Source: A.W. Oltmans, D.A. Klinefelter, and T.L. Frey, *Agricultural Financial Reporting and Analysis,* Century Communications, Inc, Niles, Ill., 1992.

Net Worth at the beginning of the period + Net Investments, gifts, or inheritances + Net Income − Withdrawals for family and gifts = Net Worth at the end of the period

Intuitively, the relationship is that net income describes all the change in net worth between two successive balance sheets, with the two exceptions of outside money coming into or withdrawals going out of the business. Stated another way, the only way to increase net income is to earn it (or inherit it). In fact, the theoretical definition of net income is the increase in net assets (i.e., net worth) from one point in time to the next.

An important task of the statement of owner equity is to segregate the effects of earned net worth change from the effects of value change. In addition to being relatively easy to interpret, the summary is also easy to calculate. The change in market value is composed of two parts: (1) change due to earnings (change in cost net worth) and (2) change due to value change. Subtracting the two change amounts as indicated will isolate the change in market value net worth resulting from inflation (or value change).

This segregation can be important to business management analysis. It is quite conceivable that market value net worth will increase from year to year while the cost net worth decreases. In other words, inflation has totally clouded the fact that more has been withdrawn from the business than has been earned. The farm or ranch may have actually been disinvesting, but that fact is overshadowed by the increases or appreciation in asset values. Many U.S. farmers and ranchers found themselves in this position during the late 1970s and early 1980s. Each year they refinanced their operations at increasingly higher debt levels. Unfortunately, those new loans were collateralized by nothing more solid than appreciating asset values.

THE INCOME STATEMENT AND THE STATEMENT OF CASH FLOWS

Earlier in this chapter, in actually a rather unlikely way and possibly an overly simplified fashion, the overall purpose of the statement of cash flows was laid out. The statement of cash flows grew out of the concern that the cash account, although reported on the balance sheet, was not given nearly enough emphasis, especially considering that cash is the lifeblood of any business. Consider a few basic questions: How can a business show accrual profits year after year and fail? How can a business accumulate accrual losses year after year and not fail? How are new loans reflected on the income statement? How is the purchase of a long-lived asset reflected on the income statement?

The answer to each of these questions lies in the difference between net income and cash flow. A business can accumulate losses and not fail (at least for several and perhaps many years) if it passes no cash out in withdrawals, bears heavy non-cash expenses (like depreciation), and/or receives outside funds. A profitable business can fail through excessive withdrawals and/or unwise, unproductive, or untimely investments. None of these answers are adequately reflected on an accrual income statement. Similarly, asset purchases and loan activity are not reflected, and the impact of many asset sales are not adequately reflected, on the income statement, because an exchange of an asset for an asset, a liability for an asset, or an asset for a liability does not meet the definition of net income. To be able to view the effects of these changes on cash position of the business, an income statement is wholly inadequate—a statement of cash flows is needed.

The structure of the statement of cash flows is relatively straightforward, as depicted in Figure 2.4. The overall net cash flow for the time period (ending cash balance from the end-of-year balance sheet) is separated into three components: operating (normal income-producing activities and associated expenses), investing (purchase or sale of long-lived assets), and financing (taking out new loans or repaying principal on old ones). Two approaches may be used for the operating section. The direct method (see Figure 2.4) takes inflows and outflows directly from a cash-based record book's receipt and expenditure summaries. The indirect method begins with accrual net income and adjusts the accrual amount for non-cash items (such as depreciation and inventory changes) to obtain a cash flow amount.

Because AFRA assumes that a cash-based record book is maintained, it employs a direct approach for the statement of cash flows (Figure 2.5). But it also presents a schedule to reconcile net income and cash flows from operating activities (Figure 2.6).

PROJECTING FINANCIAL STATEMENTS

Preparing financial statements that are projected into the future can be an extremely valuable management exercise. A complete set of projected (also called pro forma) financial statements will give a good picture of where the farm or ranch is headed financially. Having this picture before the production year begins can help to revise previous decisions concerning marketing and production plans, to determine the best plan for purchases of new or replacement capital assets, and to determine existing debt repayment capabilities for the coming year. The bottom line is that farmers or ranchers will be able to begin the year with a detailed plan or outline of where they are going; in other words, they have a road map to follow in getting from one point to another. They have some idea of the obstacles that will have to be faced as well.

In taking a long trip, travelers may change their minds about the direct route to be taken, about side excursions, and sometimes even about the final destination. Similarly, farmers or ranchers may change their minds about sidedress fertilizer rates

Cash flows from operating activities

+ Inflows
- Outflows

Cash flows from investing activities

+ Inflows
- Outflows

Cash flows from financing activities

+ Inflows
- Outflows

Net cash flows

Figure 2.4. Structure of the statement of cash flows.

STATEMENT OF CASH FLOWS

SCF

For a 12-month period Ending _____ , 19____

Actual

Projected

Name _____

Address _____ Phone _____

CASH FLOWS FROM OPERATING ACTIVITIES:

Cash received from farm operations:

Feeder livestock/poultry sales (IS, line 1a) $_____ Crops/feed (IS, line 2a) .... $_____

Livestock & poultry products (IS, line 3). _____ Custom work (IS, line 4) .... _____

Gov't payments, cash & certs (IS, line 5) _____ Patronage div. (IS, line 6a) .... _____

Hedging acc't withdrawals (Sch. 17, line b) _____ Other revenue (IS, line 9a) .... _____ + $ _____ (1)

Cash received from non-farm income and operations:

Wages (IS, line 22) .... $_____ Interest, dividends (IS, line 23a) . $_____

Royalties (IS, line 25)................ _____ Other revenue (IS, line 28a) .... _____

Cash income from other entities, farms, businesses & real estate (Sch. 25, col. a) .... _____ + _____ (2)

Cash paid for farm operating activities:

Feeder livestock/poultry (IS, line 11) ..... $_____ Feed purchases (IS, line 12) ...... $_____

Operating expenses (IS, line 14)........... _____ Interest expense (IS, line 18a) .... _____

Hedging acc't deposits (Sch. 17, line c)................ − _____ (3)

Cash expenses paid in non-farm operations (other entities, farms, businesses) (Sch. 25, col. b)............ − _____ (4)

Income and social security taxes paid in cash (IS, line 31a)................ − _____ (5)

Extraordinary items received or paid in cash (IS, line 32) .... ± _____ (6)

NET CASH INCOME (add lines 1 thru 6) .... (7) $ _____

Cash withdrawals for family living (SOE, cash portion of line 3a) .... − $ _____ (8)

Cash withdrawals for investments into personal assets (SOE, cash portion of line 3b) .... − _____ (9)

NET CASH PROVIDED BY OPERATING ACTIVITIES (line 7 minus lines 8 and 9) .... (10) $ _____

CASH FLOWS FROM INVESTING ACTIVITIES:

Cash received from the sales of:

Raised breeding & dairy livestock, not capitalized and not depreciated (Sch. 21, line b) .... + $ _____ (11)

Purchased and raised breeding & dairy livestock, capitalized and depreciated (Sch. 22, column a) .... + _____ (12)

Machinery and equipment (Sch. 23, column a)................ + _____ (13)

Farm real estate; other farm assets (Sch. 23, column a) .... + _____ (14)

Bonds and securities; investments in other entities; other non-farm assets (Sch. 24, column a):........... + _____ (15)

Cash paid to purchase:

Breeding & dairy livestock .... − _____ (16)

Machinery and equipment .... − _____ (17)

Farm real estate; other farm assets .... − _____ (18)

Capital leased assets .... − _____ (19)

Bonds and securities; investments in other entities; other non-farm assets................ − _____ (20)

NET CASH PROVIDED BY INVESTING ACTIVITIES (add lines 11 thru 20) .... (21) $ _____

CASH FLOWS FROM FINANCING ACTIVITIES:

Operating and CCC loans received (including interest paid by loan renewal) .... + $ _____ (22)

Term debt financing—loans received................ + _____ (23)

Cash received from gifts, inheritances, and paid-in capital (SOE, cash portion of line 4a)................ + _____ (24)

Personal investments of cash added into business assets (SOE, cash portion of line 4b)................ + _____ (25)

Operating debt principal payments (including repayment of CCC loans for redeemed grain) .... − _____ (26)

Term debt principal payments: Scheduled payments .... − _____ (27)

Unscheduled payments................ − _____ (28)

Principal portion of payments on capital leases .... − _____ (29)

Cash distributions of dividends, capital, or gifts (SOE, cash portion of line 5)................ − _____ (30)

NET CASH PROVIDED BY FINANCING ACTIVITIES (add lines 22 thru 30)................ (31) $ _____

NET INCREASE (DECREASE) IN CASH AND CASH EQUIVALENTS (add lines 10, 21, and 31)................ (32) $ _____

Cash and cash equivalents reported on the beginning-of-year Balance Sheet: (BS, lines 1, 2, and PIK certificates

listed as part of line 3) .... (33) $ _____

Cash and cash equivalents, as calculated, at the end of year (line 32 plus line 33) .... (34) _____

For cash reconciliation purposes, compare line 34 to the end-of-year cash and cash equivalents reported on line 35 below:

Cash and cash equivalents reported on the end-of-year Balance Sheet: (BS, lines 1, 2, and PIK certificates listed

as part of line 3) .... (35) $ _____

This form is copyrighted. It is a violation of the U.S. Copyright Law to reproduce it in any manner. To order forms, write or call Century Communications, Inc. 6201 Howard St. Niles, IL 60714 708/647-1200 or Doane Agricultural Services Co. 11700 Berman Dr. St. Louis, MO 63146. 314/569-2700

Figure 2.5. AFRA statement of cash flows.

Source: A.W. Oltmans, D.A. Klinefelter, and T.L. Frey, *Agricultural Financial Reporting and Analysis,* Century Communications, Inc, Niles, Ill., 1992.

(SCF, Part 2) RECONCILIATION OF NET INCOME TO NET CASH PROVIDED BY OPERATING ACTIVITIES

NET INCOME (IS, line 33).. (a) $_____

ADJUSTMENTS that reconcile Net Income to Net Cash Provided By Operating Activities:
REVENUE ITEMS (accrual and non-cash adjustments):
Change in feeder livestock & poultry inventory plus transfers (IS, line 1b plus 1c)....................................±$_____ (b)
Change in crops and feed inventory (IS, line 2b)..± _____ (c)
Non-cash patronage dividends (IS, line 6b).. _____ (d)
Change in farm accounts receivable (IS, line 7) ....± _____ (e)
Change in hedging account equity (Sch. 17, line a) ....± _____ (f)
Other non-cash farm revenue (IS, line 9b)... _____ (g)
Interest, dividends and capital gains earned and re-invested (IS, line 23b) .... _____ (h)
Change in accrued interest earned (IS, line 24) ....± _____ (i)
Other non-cash, non-farm income (IS, line 28b).. _____ (j)
Capital adjustment and gain (loss) on sales of: Breeding livestock and machinery (IS, line 20) ....± _____ (k)
Non-farm assets (IS, line 26) ....± _____ (l)
TOTAL REVENUE ADJUSTMENTS (add lines b thru l)..± _____ (m)
Enter the total from line (m) but REVERSE THE SIGN (i.e., if line m is positive, enter the amount here, on
line n, as a negative figure; if line m is negative, enter the amount here, on line n, as a positive figure) .... (n) $_____

EXPENSE ITEMS (accrual and non-cash adjustments):
Expense adjustments for unused assets and unpaid items (IS, line 15) ....±$_____ (o)
Depreciation (IS, line 16)... _____ (p)
Change in accrued interest payable (IS, line 18b) ....± _____ (q)
Adjustment for noncash expenses on other entities, farms, businesses & real estate (Sch. 25, col. c) .... _____ (r)
Change in accrued tax and deferred tax on current assets (IS, line 31b) ....± _____ (s)
TOTAL EXPENSE ADJUSTMENTS (Add lines o thru s) .... (t) $_____

minus CASH WITHDRAWALS FOR FAMILY LIVING (from line 8 on the front side of this form) .... (u) −_____
minus CASH WITHDRAWALS FOR INVESTMENT INTO PERSONAL ASSETS (from line 9 on the front
side of this form) .... (v) −_____

NET CASH PROVIDED BY OPERATING ACTIVITIES (add lines a, n, t, u, and v) .... (w) $_____
(Should agree with line 10 on the front side of this form.)

FOOTNOTES TO THE STATEMENT OF CASH FLOWS

This form is copyrighted. It is a violation of the U.S. Copyright Law to reproduce it in any manner. To order forms write or call Century Communications Inc., 6201 Howard St., Niles, IL 60714; 708/647-1200 or Doane Agricultural Services Co. 11701 Borman Dr. St. Louis, MO 63146; 314/569-2700.

Figure 2.6. AFRA statement of cash flows, reconciliation of net income.

Source: A.W. Oltmans, D.A. Klinefelter, and T.L. Frey, *Agricultural Financial Reporting and Analysis,* Century Communications, Inc, Niles, Ill., 1992.

or feed formulations after the production year has started. A change of mind simply implies that managers are updating or revising their plans based on new information; it does not imply that the planning process is useless. Going through the planning process will point out unrealistic goals and will help to avoid obstacles and potential problems. Knowing that a problem exists will alert the manager to find a way around it before it occurs.

The best, most useful plan or road map for a farmer or rancher is the cash flow budget because in that budget (if all of the AFRA forms and worksheets have been completed) all of the inflows and outflows of cash, feed requirements, crops produced, livestock, and other major inputs are summarized. But cash flow provides an incomplete picture of the business. An accrual income statement is also needed.

The best planning process can be summarized as an iterative (or circular) pattern that begins with the market outlook for the coming year. The steps in this pattern are:

1. Establish short-term and long-term business performance and personal goals.
2. Prepare a balance sheet, income statement, and statement of cash flows for the previous year.
3. Determine the marketing and price outlook or projections for each of the major commodities.
4. Develop a preliminary marketing plan.
5. Determine optimal levels of input and output from each commodity, based upon production function principles.[3]
6. Combine production plans for each commodity and develop an optimal whole-farm or whole-ranch plan based upon projected input and output prices and resource constraints.[4]
7. Prepare a projected whole-farm or whole-ranch budget (or preliminary projected income statement) from the production plan prepared in Step 6.
8. Prepare a projected cash flow budget based on the plan and budget developed in Steps 6 and 7.
9. Check the levels of cash borrowing requirements projected by the cash flow budget.
   a. Are the requirements reasonable in relation to borrowing capacity?
   b. Can short-term cash flow problems be averted by modifying purchasing patterns?
   c. Can short-term cash flow problems be averted by modifying the output marketing plans?
10. Prepare a projected balance sheet (cost basis), income statement, and statement of cash flows.
11. Analyze the projected financial statements and assess projected performance.
12. Revise the marketing and production plans developed in Steps 4 and 5. Repeat the planning process beginning with Step 3 based on the new marketing plan and new restrictions on borrowing.

Once a cash flow budget has been prepared after as many repetitions of this process as necessary, we can turn our attention to preparation of a complete financial statement package. It can be used to obtain financing and as a road map throughout the year. The plan can also be used at a later date as a measuring stick to test how well the management of the business actually performed as compared to plans.

After a complete cash flow budget has been prepared, the planning process continues, with a pro forma income statement and a projected statement of cash flows. But first, a projected balance sheet is needed.

Balance Sheet Projection

A projected balance sheet should be prepared for the last day of the calendar or fiscal year so that it can be directly compared to past actual statements. It is necessary to project only the cost side of the balance sheet, as the only two real goals of projecting a balance sheet are to provide all of the information needed for the income statement and to provide a check based on the statement of owner equity.

Completing a projected balance sheet is relatively easy for two reasons: all of the calculations of market values and contingent tax liabilities are avoided by ignoring the market value column; and most of the year-end inventory information has already been prepared in the crop, feed, and livestock budget sheets of the cash flow budget (see the last columns of Figures 2.7 and 2.8). Furthermore, the term debt loan payments (Figure 2.9) and capital purchases have been projected (Figure 2.10) in the

C-2

BUDGET SHEET FOR CROP PRODUCTION

For 19 _____ C-2

Crop to be Grown	Field		Yield Per Acre	Share %	Total Production	Seed			Fertilizer			Chemicals		
	No.	Acres				Variety	Rate per Acre	Total Quantity	Analysis	Rate per Acre	Total Quantity	Kind	Rate per Acre	Total Quantity

This form is copyrighted. It is a violation of the U.S. Copyright Law to reproduce it in any manner. To order forms write or call Century Communications Inc., 6201 Howard St., Niles, IL 60648; 708/647-1200

Figure 2.7. Budget sheet for crop production, AFRA Schedule C-2.

Source: A.W. Oltmans, D.A. Klinefelter, and T.L. Frey, *Agricultural Financial Reporting and Analysis,* Century Communications, Inc, Niles, Ill., 1992.

C-4   **BUDGET SHEET FOR LIVESTOCK PRODUCTION**   For 19_____   C-4

Kind	Beginning Inventory			Purchases					No. Raised	Transferred in (out)	No. Died	Sales					Ending Inventory		
	No.	Wt./Hd.	Value	Mo.	No.	Wt./Hd.	$/Unit	Cost				Mo.	No.	Wt./Hd.	$/Unit	Value	No.	Wt./Hd.	Value
FEEDER STOCK — LIVESTOCK AND POULTRY																			
			$				$	$							$	$			$
BREEDING STOCK — LIVESTOCK AND POULTRY																			
			$				$	$							$	$			$

This form is copyrighted. It is a violation of the U.S. Copyright Law to reproduce it in any manner. To order forms write or call Century Communications Inc., 6201 Howard St., Niles, IL 60714; 708/647-1200 or Doane Agricultural Services Co., 11701 Borman Dr., St. Louis, MO 63146; 314/569-2700.

Figure 2.8. Budget sheet for livestock production, AFRA Schedule C-4.
Source: A.W. Oltmans, D.A. Klinefelter, and T.L. Frey, *Agricultural Financial Reporting and Analysis,* Century Communications, Inc, Niles, Ill., 1992.

development of the cash flow budget. Beyond these schedules or budget sheets, the cash flow statement itself will state the ending cash, savings, and loan balances. The most difficult task that remains is to complete the depreciation schedule for the coming year and to estimate accrued interest liabilities (see Figure 2.11). All other balance sheet entries will normally be estimated to remain the same as they were at the beginning of the year. Modified cost values for marketable and not readily marketable securities, machinery, breeding stock, retirement accounts, personal vehicles, household assets, and real estate will remain the same except for adjustments of purchases, sales, and depreciation.

Income Statement Projection

Once the year-end projected balance sheet has been prepared, a projected income statement can be completed very quickly. Use the actual beginning balance, projected revenues, and expenditures from the cash flow budget and the projected ending balance sheet. No additional information is needed to prepare the pro forma income statement.

C-7   TERM DEBT LOAN PAYMENTS   For 19_____   C-7

Beginning Balance on Previous Loans	Due Date	Interest Rate	Payments	Jan.	Feb.	March	April	May	June	July	August	Sept.	Oct.	Nov.	Dec.	Ending Principal Balance
$			Principal													$
/////////			Interest													/////////
/////////																/////////
/////////																/////////
/////////																/////////
/////////																/////////
/////////																/////////
/////////																/////////
/////////																/////////
/////////																/////////
/////////																/////////
NEW LOANS																
/////////																/////////
/////////																/////////
/////////																/////////
/////////																/////////

This form is copyrighted. It is a violation of the U.S. Copyright Law to reproduce it in any manner. To order forms write or call Century Communications Inc., 6201 Howard St., Niles, IL 60714; 708/647-1200 or Doane Agricultural Services Co., 11701 Borman Dr., St. Louis, MO 63146; 314/569-2700.

Figure 2.9. Term debt loan payments, AFRA Schedule C-7.
Source: A.W. Oltmans, D.A. Klinefelter, and T.L. Frey, *Agricultural Financial Reporting and Analysis,* Century Communications, Inc, Niles, Ill., 1992.

Statement of Cash Flows Projection

Projecting the statement of cash flows, if a cash flow budget, projected balance sheet, and projected income statement have already been completed, is a mechanical process. Beginning and ending cash balances come from the actual and projected balance sheets, cash flow items come from the cash flow budgets, and net income and accrual adjustments come from the income statement.

Statement of Owner Equity Projection

The final step in the projection process is to prepare the statement of owner equity as discussed earlier in this chapter. This statement should be prepared to double-check the accuracy of the figures.

C-6

CAPITAL EXPENDITURES AND REPAIRS For 19_____ C-6

Description	Number or Quantity	Month	Purchases & Capital Repairs		Operating Repairs	
			Machinery and Equipment	Buildings and Improvements	Machinery and Equipment	Buildings and Improvements
			$	$	$	$

This form is copyrighted. It is a violation of the U.S. Copyright Law to reproduce it in any manner. To order forms write or call Century Communications Inc., 6201 Howard St., Niles, IL 60714, 708/647-1200 or Doane Agricultural Services Co., 11701 Borman Dr., St. Louis, MO 63146, 314/569-2700

Figure 2.10. Capital expenditures and repairs, AFRA Schedule C-6.
Source: A.W. Oltmans, D.A. Klinefelter, and T.L. Frey, *Agricultural Financial Reporting and Analysis,* Century Communications, Inc, Niles, Ill., 1992.

SUMMARY

Although each of the major financial statements has a unique function and assumptions very different from its counterparts, all are closely tied together and represent different views of the same business story.

All financial transactions throughout the year can be analyzed in an A = L + NW format including income earned and expenses incurred since income and expense accounts are subsets of the net worth account. A summary of these accounts can thus provide an income statement. And yet, if filtered through a trial balance at any point in time, these same accounts would provide a balance sheet. If non-cash transactions are ignored and only the transactions with an entry in the cash account are totaled, the cash flow statement would result.

The balance sheet summary and statement of owner equity show many of these relationships and also show the effect of value change relative to earned change. A very important relationship that falls out very quickly is:

SCHEDULE 11 — NOTES & CONTRACTS RECEIVABLE

To Whom	Orig. Date	Purpose and/or Security	Due Date	Interest Rate	Payment Dates	Current Principal Balance	Accrued Interest	Portion of Principal Due within 12 months	Due beyond 12 months
NOTES									
						$	$	$	$
CONTRACT SALES									
					TOTAL	$		$	$

SCHEDULE 12 — NON-CURRENT LIABILITIES

To Whom	Orig. Date	Purpose and/or Security	Due Date	Interest Rate	Payment Dates	Current Principal Balance	Accrued Interest	Portion of Principal Due within 12 months	Due beyond 12 months
NOTES									
						$	$	$	$
						(a) Sub-Total			$
CAPITAL LEASES									
						$	$	$	$
						(b) Sub-Total			$
SALES/LAND CONTRACTS									
						$	$	$	$
						(c) Sub-Total			$
REAL ESTATE MORTGAGES									
						$	$	$	$
						(d) Sub-Total			$
OTHER LOANS									
						$	$	$	$
						(e) Sub-Total			$
					(f) TOTAL for entire schedule	$		$	$

Figure 2.11. Noncurrent liabilities, AFRA Schedule 12.

Source: A.W. Oltmans, D.A. Klinefelter, and T.L. Frey, *Agricultural Financial Reporting and Analysis,* Century Communications, Inc, Niles, Ill., 1992.

This form is copyrighted. It is a violation of the U.S. Copyright Law to reproduce it in any manner. To order forms write or call Century Communications Inc. 6201 Howard St., Niles, IL 60714. 708/647-1200 or Doane Agricultural Services Co. 11770 Borman Dr. St. Louis, MO 63146. 314/569-2700

Net Worth (beginning) + Net Investments + Net Income − Withdrawals = Net Worth (ending)

This relationship drives home the fact that the only way to increase net worth (with the exception of gifts and inheritances) is to earn it through a positive net income.

An iterative planning sequence is necessary to fully develop a comprehensive logical forward plan. In brief, the steps of the planning process include the following:

1. Goals
2. Prepare past financial statements
3. Price projections
4. Marketing plan
5. Optimal combination of inputs and outputs
6. Optimal whole-farm plan
7. Projected whole-farm budget
8. Projected cash flow budget
9. Assess borrowing requirements
10. Project financial statements
11. Analyze projected financial statements
12. Revise marketing plan and other projections

NOTES

1. For a further justification of the need for double-column balance sheets, see A.W. Oltmans, D.A. Klinefelter, and T.L. Frey, pp. 10 and 19, and the report of the Farm Financial Standards Task Force.

2. See Financial Accounting Standards Board, *Statement of Financial Accounting Standards No. 95—Statement of Cash Flows,* FASB, Norwalk, Conn., 1992.

3. An excellent source of further background material concerning production economics principles can be found in R.D. Kay and W.E. Edwards, *Farm Management: Planning, Control and Implementation,* 3rd. ed., McGraw-Hill, New York, N.Y., 1993. Read especially Chapters 7 and 8.

4. Ibid. Chapters 10 and 11 present one means of obtaining the optimal whole-farm plan. Linear programming may also be used as can programmed budgeting.

RECOMMENDED READINGS

J. Bickers, "How to Take Your Financial Temperature," *Progressive Farmer,* September 1982, p. 17.

Ronald D. Kay and William E. Edwards, *Farm Management: Planning, Control, and Implementation,* 3rd, ed., McGraw-Hill, New York, N.Y., 1993.

John B. Penson, Jr., and Clair J. Nixon, *Understanding Financial Statements in Agriculture,* Agri-Information Corporation, College Station, Tex., 1983.

James D. Libbin and Lowell B. Catlett, *Farm and Ranch Financial Records,* Macmillan, New York, N.Y., 1987.

CHAPTER 3

A Cash Flow Plan

A comprehensive business plan will contain all of the components (or steps) listed in Chapter 2; it will be complete in the sense that all predictable events will have been analyzed and it will be consistent in the sense that all components will have been included and will carry through from one report to the other. We must begin our discussion of the development of a comprehensive business plan by taking stock of the current position of the business, noting the resources available to the business; the limitations forced by weather, soil, and water and human resources; and the goals of the owner and/or operator. The situation facing the business can be changed, or at least most of its parts can be changed, but those conditions set the stage for what we have to work from. The changes or directions desired then must be formulated and formalized into plans, activities, and processes that will move us from where we are to where we want to be. Those plans, which recognize our starting conditions, must be feasible; that is, they must be possible and must live within or be consistent with our situation and our limitations. The business plan must, for example, be consistent with the requirement that a positive bank balance be maintained and with the limitations imposed by credit availability. A crop plan also must be consistent with achievable yields. Once the plan is developed, its consequences, both in terms of accrual accounting principles and of the ending position, must be evaluated. In short, a plan builds upon the beginning position and is assessed by its feasibility, which can be judged by acceptance of limitations and by its consequences.

A CASE STUDY: SNAKE BELLY FARMS

To see how all of the components of a comprehensive, consistent cash flow plan hang together, let's leave the conceptual discussion and jump into an actual plan. First, we must establish the limitations and position of the business. Joe Farmer operates Snake Belly Farms, a 720-acre operation that has the capability of producing a variety of crops, including corn, wheat, cotton, soybeans, and alfalfa. Joe has experience in producing each of these crops and has adequate machinery for each. He has no experience in producing higher-valued vegetable or fruit crops, although he would like to experiment with alternative crops sometime in the future; however, at the present time he lacks the machinery, experience, and capital base to produce alternative crops. Snake Belly Farms contains 400 acres of Class I land that can be used for any crop; 200 acres of Class II land that can be planted only to drilled crops (non-row crops) such as wheat, alfalfa, or pasture; 100 acres of Class III land that are not suitable for cropping other than permanent pasture; and 20 acres committed to roads, ditches, and farmstead. To make use of the pastures, the previous owner fenced the farm and built facilities to maintain a cattle herd—either a cow-calf operation or a stocker operation would work without additional facilities.

Joe Farmer bought Snake Belly Farms in 19X1 on a land contract with the previous owner. Although he owns an extensive line of machinery in excellent condition, his expansion capabilities are limited by his leveraged financial position.

Figure 3.1 shows a rough diagram of the physical layout of Snake Belly farms, including field and building locations. Figure 3.2 presents a complete balance sheet with supporting schedules for Snake Belly Farms as of December 31, 19X1. The initial balance sheet is perhaps the most crucial document in the planning process. It is the necessary beginning step, because it describes the initial position of the business

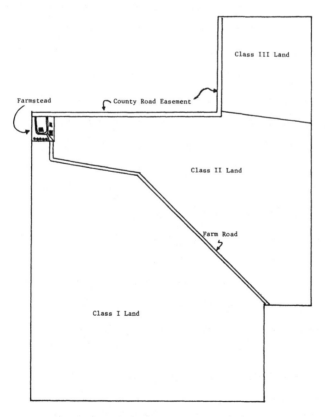

Figure 3.1. Layout, Snake Belly Farms.

and defines the physical and financial resources and limitations into which the plan must fit.

From the real estate schedule (Figure 3.2c), Snake Belly Farms has 720 acres spread across three land classes and a comprehensive machinery complement is owned (Figure 3.2e). Figure 3.2d lists the noncurrent financial obligations of Snake Belly Farms. Since the farm was purchased recently, substantial debt obligations exist. These obligations will play a major role in the shape of the eventual cash flow plan as they represent a heavy drain on the cash resources of the business.

The overall financial position of Snake Belly Farms on December 31, 19X1, is best shown through a rudimentary financial ratio analysis of the balance sheet (Figure 3.2a), as shown in Table 3.1.[1,2] Low current ratio indicates a tight liquidity position. Most financial analysts would look for a current ratio of about 1.5 while lenders might prefer a value of 2 or higher.

BALANCE SHEET
Snake Belly Farms
December 31, 19X1

ASSETS			LIABILITIES		
CURRENT ASSETS	COST	MARKET VALUE	CURRENT LIABILITIES		MARKET VALUE
Cash on hand and in checking accounts	10,000	10,000	Accounts payable		7,000
Feeder livestock & poultry	192,456	192,456	Notes payable		165,000
Crops and feed	12,800	12,800	Principal due within 12 months		
Investment in growing crops	40,800	40,800	on all non-current liabilities		41,433
Supplies	2,375	2,375	Accrued interest on:		
			Accounts 0 Notes	4,936	
			Non-currrent liabilities	4,234	9,170
			Accrued tax liability:		
			Real estate		1,625
TOTAL CURRENT ASSETS	258,431	258,431	TOTAL CURRENT LIABILITIES		224,228
NON-CURRENT ASSETS			NON-CURRENT LIABILITIES		
Machinery, equipment, trucks		163,475	(Principal due beyond 12 months)		
Cost or basis	368,950		Notes payable		945,562
Less accumulated dep	332,410	36,540			
Real estate		914,000			
Cost or basis	1,000,000		TOTAL NON-CURRENT LIABILITIES		945,562
Less accumulated dep	49,500	950,500			

			TOTAL CURRENT AND NON-CURRENT LIABILITIES	COST	MARKET VALUE
TOTAL NON-CURRENT ASSETS	987,040	1,077,475		1,169,790	1,169,790
TOTAL BUSINESS ASSETS	1,245,471	1,335,906	Deferred tax on non-current assets		42,272
Personal Assets		21,500	TOTAL LIABILITIES	1,169,790	1,212,062
			Owner equity:		
			Retained Earnings	55,681	55,681
			Contributed Capital	20,000	20,000
			Personal Net Worth	--	21,500
			Valuation Equity	--	48,163
TOTAL ASSETS	1,245,471	1,357,406	TOTAL OWNER EQUITY	75,681	145,344
			TOTAL LIABILITIES & OWNER EQUITY	1,245,471	1,357,406

A. December 31, 19X1, projected balance sheet.

Figure 3.2. Projected balance sheet, Snake Belly Farms, December 31, 19X1.

The four solvency ratios all tell basically the same story: a very highly leveraged, and consequently risky, financial position. This result was expected given that the farm was purchased in 19X1 with 100 percent debt financing. Land devaluation has eroded the market value of fixed assets and owner equity as well.

The profitability ratios are surprisingly quite good. Despite what some might think of as a low net farm income in 19X1 ($19,000), rates of return on capital exceeded certificate of deposit rates common in 19X1.

In summary, the financial position of Snake Belly Farms is tenuous. The liquidity and solvency ratios indicate a very highly leveraged, risky financial picture. But they certainly do *not* indicate that the business is in severe trouble. Profitability ratios indicate that adequate returns are being made. One major problem in prices or yields, however, could move Snake Belly Farms into a very dangerous financial position.

DESCRIPTION	QUANTITY	UNIT	VALUE
BALING TWINE	40	ROLLS	1,000
DIESEL FUEL	500	GALS	500
MOTOR OIL	25	GALS	100
MISCELLANEOUS			775
		TOTAL	2,375

B. Supplies schedule.

	% OWN	DATE ACQ	COST OR BASIS	ACC DEP	ADJUSTED VALUE	MKT VALUE
LAND (400 ACRES, CLASS I)	100	12/86	440,000	0	440,000	400,000
RESIDENCE	100	12/86	50,000	0	50,000	55,000
SERVICE BLDG	100	12/86	90,000	13,500	76,500	80,000
IMPROVEMENTS	100	12/86	160,000	24,000	136,000	150,000
LAND (200 ACRES, CLASS II)	100	12/86	120,000	0	120,000	100,000
IMPROVEMENTS	100	12/86	80,000	12,000	68,000	75,000
LAND (120 ACRES, CLASS III)	100	12/86	60,000	0	60,000	54,000
ACRES 720	TOTALS		1,000,000	49,500	950,500	914,000

C. Real estate schedule.

TO WHOM	PURPOSE OR SECURITY	DUE DATE	INT RATE	PAY DATE	CURR PRIN BALANCE	PART DUE (12 MOS)	PRIN DUE 12 MOS	ACCRUED INT
1ST NATL	TON BALER	12/01/X3	14.00	12/01	43,956	12,780	31,176	505
1ST NATL	3 PUMPS	12/15/X3	13.00	12/15	7,114	3,340	3,774	40
PREVIOUS OWNER	FARM	12/15/Y7	9.00	12/15	935,925	25,313	910,612	3,689
	TOTALS				986,995	41,433	945,562	4,234

D. Noncurrent debt schedule.

Figure 3.2. (*continued*).

ITEM	YEAR, MAKE AND MODEL	DATE ACQ	% OWN	COST OR BASIS	ACC DEP	ADJUSTED COST	MARKET VALUE
65 HP TRACTOR		1963	100	5,500	5,500	0	1,400
96 HP TRACTOR		1971	100	10,300	10,300	0	4,300
130 HP TRACTOR		1984	100	43,600	43,600	0	28,000
2-ROW COTTON PICKER		1985	100	77,000	77,000	0	27,500
14 FT. SWATHER		1986	100	29,000	22,910	6,090	9,200
1-TON BALER		1987	100	65,000	37,700	27,300	27,000
4-ROW PLANTER		1985	100	6,800	6,800	0	3,400
4-ROW CULTIVATOR		1983	100	3,000	3,000	0	1,500
14 FT. OFFSET DISK		1983	100	11,500	11,500	0	5,750
13 FT. DRILL		1985	100	4,300	4,300	0	2,150
12 FT. PLANE		1983	100	5,000	5,000	0	2,500
14 FT. FLOAT		1983	100	900	900	0	600
4-ROW LISTER		1983	100	3,200	3,200	0	1,600
4-16 IN. MOLDBOARD PLOW		1983	100	8,000	8,000	0	4,000
4-ROW SHREDDER		1983	100	5,000	5,000	0	2,500
8 COTTON TRAILERS		1980	100	19,200	19,200	0	9,600
12 FT. SPRAYER		1983	100	2,750	2,750	0	1,375
FRONT-END LOADER (ATTACH)		1983	100	5,100	5,100	0	2,550
SEMI-TRUCK & FLATBED		1985	100	30,000	30,000	0	15,000
RAKES		1983	100	6,800	6,800	0	3,400
V-DITCHER		1983	100	2,500	2,500	0	1,250
3 NATURAL GAS PUMPS		1986	100	15,000	11,850	3,150	7,500
PICKUP	1978 FORD	1978	100	7,000	7,000	0	1,000
PICKUP	1971 FORD	1971	100	2,500	2,500	0	400
		TOTALS		368,950	332,410	36,540	163,475

E. Machinery schedule.

ITEM	QUANTITY	UNIT	$/UNIT	VALUE
CORN	2,000	BU	2.15	500
ALFALFA	100	TONS	85.00	100
				775
			TOTAL	2,375

F. Crops and feed schedule.

DESCRIPTION	NUMBER	AVERAGE WEIGHT	$/CWT	VALUE
STOCKER STEERS	540	540	66.00	192,456
			TOTAL	192,456

G. Feeder livestock schedule.

CROP	ACRES	$/ACRE	VALUE
ALFALFA I	300	100.00	30,000
PASTURE II	100	50.00	5,000
PASTURE III	100	40.00	4,000
ALFALFA II	20	90.00	1,800
		TOTAL	40,800

H. Investment in growing crops schedule.

Figure 3.2. (continued).

DESCRIPTION	AMOUNT
FEED STORE	1,000
FUEL DISTRIBUTOR	3,000
UTILITY COMPANIES	3,000
TOTAL	7,000

I. Accounts payable schedule.

TO WHOM	PURPOSE OR SECURITY	DUE DATE	INT RATE	PAY DATE	PRINCIPAL BALANCE	ACCRUED INT
1ST NATL	CATTLE	06/01/X2	12.00	10/01	165,000	4,936
		TOTAL			165,000	4,936

J. Notes due within 12-months schedule.

DESCRIPTION	VALUE
SUBARU (1981 GL)	2,200
MG (1979 B)	2,800
HOUSEHOLD GOODS	12,000
UNIVERSAL LIFE INSURANCE POLICY (MUTUAL; 100,000 FACE VALUE)	4,500
TOTAL	21,500

K. Personal assets schedule.

Section A - Deferred Tax Estimate on Current Assets

Feeder livestock and poultry	192,456		
(minus) Purchase cost of feeders on hand	-170,100	22,356	
Crops and feed		12,800	
Supplies		2,375	
Investment in growing crops		40,800	
Total current assets that could be taxed			78,331
Farm accounts payable		7,000	
Estimated accrued interest		9,170	
Estimated accrued tax liability		1,625	
Total current liabilities that could be deducted			-17,795
NET TAXABLE CURRENT ASSETS			60,536

Section B - Deferred Tax Estimate on Non-Current Assets

		Cost		Market Value		Taxable Gain
Machinery	-	36,540	+	163,475	+	126,935
Farm real estate	-	950,500	+	914,000	-	36,500
NET TAXABLE GAINS						90,435
NET TAXABLE INCOME						150,971
Ordinary tax rate					X	28%
DEFERRED TAX						42,272

L. Deferred tax liability worksheet.

Figure 3.2. (continued).

Table 3.1. Financial ratio analysis of Snake Belly Farms, December 31, 19X1

Ratio	Definition	Value for year 19X1
Liquidity Ratios		
1. Current Ratio	Current assets/Current liabilities	258,431 / 224,228 - 1.15
2. Working Capital	Current assets - Current liabilities	258,431 - 224,228 = 34,203
Solvency Ratios		
3. Debt/Asset Ratio	Total liabilities / Total assets	1,212,062 / 1,357,406 = 0.89
4. Equity/Asset Ratio	Net worth / Total assets	145,344 / 1,357,406 = 0.11
5. Leverage Ratio	Total liabilities / Net worth	1,1212,062 / 145,344 = 8.34
Profitability Ratios		
6. Rate of Return on Farm Asssets	(Return to total capital/Total assets) x 100%	117,683 / 1,357,406 = 8.64%
7. Rate of Return on Equity	(Return to equity capital/Net worth) x 100%	11,727 / 145,344 = 8.07%
8. Operating Profit Margin Ratio	(Net farm income + interest expense - value of operator & unpaid family labor & management)/Gross revenue	(29,727 + 105,956 - 18,000)/509,611 = 23.09%
9. Net Farm Income	From income statement	29,727
Repayment Capacity		
10. Term Debt and Capital Lease Coverage Ratio	(Net farm income + Total nonfarm income + Depreciation expense + Interest on term debt and capital leases + Total income tax expense - Withdrawal for family living)/ (Annual scheduled principal and interest payments on term debt and capital leases)	(29,727 + 14,400 + 40,104 + 105,956 - 1,000 - 18,000)/(206,433 + 104,512) = 0.55
11. Capital Replacement and Debt Repayment Margin	Net farm income + Total nonfarm income + Depreciation expense - Total income tax expense - Family living withdrawals - Payment on unpaid operating debt from a prior period - Principal payments on current portion of debt and capital leases - Total annual payments on personal liabilities (if not included on withdrawals for family living)	(29,727 + 14,400 + 40,104 = 1000 - 18,000 - 7,000 - 206,433) = -35,246
Financial Efficiency		
12. Asset Turnover Ratio	Gross revenue/Average total farm assets	509,611 / 1,362,785 = 0.37%
13. Operational Ratios		
a) Operating Expense Ratio	Operating expense/Gross revenue	155,980 / 509,611 = 30.61%
b) Depreciation Expense Ratio	Depreciation expense/Gross revenue	40,104 / 509,611 = 7.87%
c) Interest Expense Ratio	Interest expense/Gross revenue	105,956 / 509,611 = 20.79%
d) Net Farm Income from Operations Ratio	Net farm income/Gross revenue	29,727 / 509,611 = 5.83%

THE PLANNING PROCESS

At the risk of over-repetition, let's review the twelve steps of the planning process outlined in Chapter 2 and relate those steps to the plans that Joe Farmer must develop for Snake Belly Farms.

1. Goals
2. Prepare past financial statements
3. Price projections
4. Marketing plan
5. Optimal combination of inputs and outputs
6. Optimal whole-farm plan
7. Projected whole-farm budget
8. Projected cash flow budget
9. Assess borrowing requirements
10. Project financial statements
11. Analyze projected financial statements
12. Revise marketing plan and other projections

Joe Farmer developed a whole-farm plan (without regard to market or production risks) that is consistent with the twelve-step planning process outlined above. To develop the basis for the whole-farm plan, Joe assessed his short-run and long-run goals and current physical and financial position, and then developed an initial marketing plan for each potential commodity, and projected input needs and yields (these projections are shown in Table 3.2). Joe next compiled all of his projections

Table 3.2. Projected 19X1 output price, purchased input, and yield projections, Snake Belly Farms

	Unit	Projected Price[a]	Projected Purchased Inputs	Projected Yield
Crops		$/unit	$/acre	Units/Acre
Alfalfa I	Ton		185	7
January		85.00		
February		81.25		
March		81.00		
June		85.00		
July		80.00		
August		82.50		
September		83.00		
October		89.25		
November		89.00		
December		91.00		
Alfalfa II	Ton[b]		175	5
Corn	Bushel			
January		2.15		
February		2.20		
March		2.25		
October		2.10		
November		2.05		
December		2.00		
Cotton	Pound[c]	.60	200	750
Pasture II	AUM[d]		30	15
Pasture III	AUM[d]		30	12.7
Soybeans	Bushel	5.40	115	45
Wheat	Bushel		131	70
July		2.25		
August		2.30		
Cottonseed	Pound	.03	0	1200[c]
			$/head	weight/head
Livestock				
Purchase Stockers	Pound	.70	13.50	450
Sell Stockers	Pound[e]	.66		740
Beef cow (calves)	Pound[f]	.70	25.00	450

[a]Season average projected price
[b]Price projects same as Alfalfa I
[c]Lint. See yield = 1.6 × lint yield
[d]Not saleable
[e]Death loss of 1 percent
[f]Calving percentage of 88 percent

and developed a whole-farm plan using a programmed budgeting and linear programming approach. The plan includes 320 acres of alfalfa, 100 acres of cotton, 200 acres of grass pasture, 80 acres of wheat, and 540 head of stocker steers as shown in Table 3.3.

From this whole-farm plan, Joe was then able to develop the cash flow budget presented in Figure 3.3. Using the financial statement projection principles discussed in Chapter 2, Joe then completed his financial statement package by preparing the projected income statement shown in Figure 3.4, the projected balance sheet shown in Figure 3.5, and the projected statement of cash flows shown in Figure 3.6.

PLAN ADEQUACY

The plan developed by Joe Farmer meets all of the existing debt repayment and cash family living obligations of the farm and family. It also accounts for crop rotational requirements, the cash obligations of growing each crop and livestock commodity, short-term financing of the production activities, repayment of short-term financing, and timing of crop and livestock sales. Joe needs to borrow $50,000 in short-term capital through an operating line of credit. Crop operating funds, along with the cattle purchase loan of 19X2, will be repaid with proceeds from crop and livestock sales by June. A cattle purchase loan for 19X2 for $155,000 will also be needed. Joe plans to sell or feed his carry-in inventory of alfalfa and corn by March and sell his 19X2 crop alfalfa in October and December. However, to provide winter feed, he plans to carry over 100 tons of his alfalfa production to 19X2. To maximize

Table 3.3. 19X1 Whole-farm plan, Snake Belly Farms

Crop	Acres		
	Class II	Class III	Class III
Alfalfa I	300		
Alfalfa II		20	
Corn	0		
Cotton	100		
Pasture II		100	
Pasture III			100
Soybeans		80	
Wheat	--	--	--
Total	400	200	100

Livestock	Head
Stocker Steers	540
Beef Cows	0

CASH FLOW BUDGET
Snake Belly Farms
Projected for 19X2

DESCRIPTION	LAST YEAR	TOTAL	JAN	FEB	MAR	APRIL	MAY	JUNE	JULY	AUG	SEPT	OCT	NOV	DEC
BEG. CASH BALANCE	0	10000	10000	4217	1000	1000	1000	1000	8707	1000	1000	1000	18200	13270
OPERATING RECEIPTS:														
CROPS AND FEED	0	242717	6885	0	733	0	0	0	0	9338	0	91213	0	134548
LIVESTOCK & POULTRY	0	247752	0	0	0	0	0	247752	0	0	0	0	0	0
PRODUCTS (LIVESTOCK)	0	0	0	0	0	0	0	0	0	0	0	0	0	0
CUSTOM WORK	0	0	0	0	0	0	0	0	0	0	0	0	0	0
GOVERNMENT PAYMENTS	0	23392	0	0	0	6109	0	0	4872	0	3248	0	0	9163
HEDGING ACCOUNT W/D	0	0	0	0	0	0	0	0	0	0	0	0	0	0
CAPITAL RECEIPTS:														
BREEDING STOCK	0	0	0	0	0	0	0	0	0	0	0	0	0	0
MACHINERY & EQUIP.	0	0	0	0	0	0	0	0	0	0	0	0	0	0
NON-FARM INCOME:														
OFF-FARM WAGES	0	14400	1200	1200	1200	1200	1200	1200	1200	1200	1200	1200	1200	1200
INTEREST AND DIVIDENDS	0	0	0	0	0	0	0	0	0	0	0	0	0	0
TOTAL CASH AVAILABLE	0	538261	18065	5417	2933	8309	2200	249952	14779	11558	5448	93413	19400	158181
OPERATING EXPENSES:														
LABOR HIRED	0	12900	850	850	850	850	850	2200	850	850	850	850	850	2200
REPAIRS-MACH & EQUIP.	0	9120	760	760	760	760	760	760	760	760	760	760	760	760
REPAIRS-BUILD/IMPROV.	0	2578	75	75	75	75	75	1075	753	75	75	75	75	75
RENTS & LEASES	0	0	0	0	0	0	0	0	0	0	0	0	0	0
SEED	0	8591	0	0	0	600	2431	0	0	5560	0	0	0	0
FERTILIZER & LIME	0	18807	0	0	0	1350	0	0	0	17457	0	0	0	0
CHEMICALS	0	12849	0	0	0	6400	675	1920	2625	533	696	0	0	0
CUSTOM MACHINE HIRE	0	9387	0	0	0	0	0	3031	0	0	0	0	0	6356
SUPPLIES	0	1500	125	125	125	125	125	125	125	125	125	125	125	125
LIVESTOCK EXPENSE	0	5400	780	600	600	600	600	600	0	0	0	420	600	600
GAS, FUEL, OIL	0	58018	0	0	0	7891	8216	8704	9660	10930	10406	511	1700	0
STORAGE/CUSTOM DRY	0	0	0	0	0	0	0	0	0	0	0	0	0	0
TAXES (REAL EST, PP)	0	2112	0	2112	0	0	0	0	0	0	0	0	0	0
INSURANCE(PROP, LIAB)	0	3300	800	0	1000	1500	0	0	0	0	0	0	0	0
UTILITIES(ELECT/GAS)	0	2040	170	170	170	170	170	170	170	170	170	170	170	170
MARKET/TRANSPORT EXP	0	0	0	0	0	0	0	0	0	0	0	0	0	0
AUTO (FARM SHARE)	0	4200	350	350	350	350	350	350	350	350	350	350	350	350
ACCOUNTS PAYABLE	0	7000	7000	0	0	0	0	0	0	0	0	-0	0	0
EDUC, TRG, & MISC.	0	4983	0	200	1233	350	1200	800	0	1200	0	0	0	0
TOTAL CASH OPER EXPS	0	162783	10910	5242	5163	21021	15452	19735	15293	38010	13432	3261	4630	10636

Figure 3.3. 19X2 projected cash flow budget, Snake Belly Farms.

CASH FLOW BUDGET
Snake Belly Farms
Projected for 19X2

DESCRIPTION	LAST YEAR	TOTAL	JAN	FEB	MARCH	APRIL	MAY	JUNE	JULY	AUG	SEPT	OCT	NOV	DEC
STOCK & FEED PURCH:														
FEEDER LIVESTOCK	0	173250	0	0	0	0	0	0	0	0	0	173250	0	0
BREEDING LIVESTOCK	0	0	0	0	0	0	0	0	0	0	0	0	0	0
FEED PURCHASED	0	4594	1458	0	0	0	0	0	0	0	0	0	3136	0
CAPITAL EXPENDITURES														
MACHINERY & EQUIP	0	17000	0	0	2000	0	0	0	0	0	0	15000	0	0
BUILDINGS & IMPROVE.	0	3710	0	0	0	0	0	0	3710	0	0	0	0	0
OTHER EXPENDITURES:														
HEDGING ACCT DEPOST	0	0	0	0	0	0	0	0	0	0	0	0	0	0
GROSS FAMILY LIV W/D	0	18000	1500	1500	1500	1500	1500	1500	1500	1500	1500	1500	1500	1500
NON-FARM BUS/INVEST	0	0	0	0	0	0	0	0	0	0	0	0	0	0
INCOME TAX & SOC SEC	0	1000	0	0	1000	0	0	0	0	0	0	0	0	0
	0	0	0	0	0	0	0	0	0	0	0	0	0	0
LOAN PAYMENTS - PRIN	0	206433	0	0	0	0	0	165000	0	0	0	0	0	41433
LOAN PAYMENTS - INT	0	104512	0	0	0	0	0	13200	0	0	0	0	0	91312
TOTAL CASH REQUIRED	0	691284	13868	6742	9663	22521	16952	199435	20503	39510	14932	196147	6130	144881
CASH AVAIL - CASH REQ		-153023	4217	-1325	-6730	-14212	-14752	50517	-5724	-27972	-9484	-102734	13270	13300
INFLOWS FROM SAVINGS		0	0	0	0	0	0	0	0	0	0	0	0	0
CASH POS BEFORE BORR		-153023	4217	-1325	-6730	-14212	-14752	50517	-5724	-27972	-9484	-102734	13270	13300
MONEY TO BE BORROWED														
-OPERATING LOANS		87576	0	2325	7730	15212	16129	0	6724	28972	10484	0	0	0
-INT & L/T LOANS		168000	0	0	0	0	0	0	0	0	0	168000	0	0
OP LOAN PAY - PRIN		87576	0	0	0	0	0	41396	0	0	0	46180	0	0
-INTEREST		1677	0	0	0	0	377	414	0	0	0	886	0	0
OUTFLOWS TO SAVINGS		0	0	0	0	0	0	0	0	0	0	0	0	0
ENDING CASH BALANCE		13300	4217	1000	1000	1000	1000	8707	1000	1000	1000	18200	13270	13300
LOAN BALANCES:														
CURR INTEREST RATE	12.00													
CURRENT YR'S OP LOAN			0	2325	10055	25267	41396	0	6724	35696	46180	0	0	0
-ACCRUED INTEREST		1677	0	23	101	253	414	0	67	357	462	0	0	0
PREV YR'S OPER LOANS			0	0	0	0	0	0	0	0	0	0	0	0
-ACCRUED INTEREST			0	0	0	0	0	0	0	0	0	0	0	0
INT & LONG TERM LOAN	1151995		1151995	1151995	1151995	1151995	1151995	986995	986995	986995	986995	1154995	1154995	1113562
TOTAL LOANS			1151995	1151995	1151995	1151995	1151995	986995	993719	1022691	1033175	1154995	1154995	1113562
CONSISTENCY CHECK:														
TOTAL INFLOWS			18085	7742	10663	23521	18329	249952	21503	40510	15932	261413	19400	158181
TOTAL OUTFLOWS			18085	7742	10663	23521	18329	249952	21503	40510	15932	261413	19400	158181
BUDGETING ERROR			0	0	0	0	0	0	0	0	0	0	0	0

Figure 3.3. (continued)

INCOME STATEMENT
Snake Belly Farms
Projected for 19X2

REVENUE			
Feeder Livestock & poultry:			
Cash sales	247,752		
Inventory change	0		
Feeder livestock transferred to breeding herd	0	247,752	
Crops and feed:			
Cash sales	242,717		
Inventory change	-4,250	238,467	
Government payments		23,392	
Gross revenue			509,611
minus feeder livestock & poultry purchases		173,250	
minus feed purchased		4,594	
VALUE OF FARM PRODUCTION			331,767
EXPENSES			
Cash operating expenses		162,785	
Accrual expense adjustments (unused assets & unpaid items)		-6,785	
Depreciation: Machinery and equipment	23,604		
Fixed farm improvements	16,500	40,104	
Total operating expenses			196,084
Income from farm operations			135,683
minus Interest expense			105,956
NET FARM INCOME FROM OPERATIONS			29,727
Capital adjusment, gain (loss)		0	
NET FARM INCOME			29,727
NON FARM INCOME			
Spouse's wage off farm		14,400	
NON-FARM INCOME			14,400
INCOME BEFORE INCOME TAXES AND EXTRAORDINARY ITEMS			44,127
Income and social security taxes			1,000
Income before extraordinary items			43,127
Extraordinary items (explain)			0
NET INCOME			43,127

Figure 3.4. 19X2 projected income statement, Snake Belly Farms.

BALANCE SHEET
Snake Belly Farms
December 31, 19X2

ASSETS			LIABILITIES		
CURRENT ASSETS	COST	MARKET VALUE	CURRENT LIABILITIES		MARKET VALUE
Cash on hand and in checking accounts	13,303	13,303	Notes payable		155,000
Feeder livestock & poultry	192,456	192,456	Principal due within 12 months		
Crops and feed	8,550	8,550	on all non-current liabilities		48,695
Investment in growing crops	40,800	40,800	Accrued interest on:		
Supplies	2,180	2,180	Accounts 0 Notes 4,623		
			Non-current liabilities 4,314		8,937
			Accrued tax liability:		
			Real estate		1,625
TOTAL CURRENT ASSETS	257,289	257,289	TOTAL CURRENT LIABILITIES		214,257
NON-CURRENT ASSETS			NON-CURRENT LIABILITIES		
Machinery, equipment, trucks		163,475	(Principal due beyond 12 months)		
Cost or basis	383,950		Notes payable		909,867
Less accumulated dep	356,014	27,936			
Real estate		914,000	TOTAL NON-CURRENT LIABILITIES		909,867
Cost or basis	1,000,000				
Less accumulated dep	66,000	934,000			
			TOTAL CURRENT AND NON-CURRENT		MARKET
			LIABILITIES	COST	VALUE
TOTAL NON-CURRENT ASSETS	961,936	1,088,975		1,124,124	1,124,124
TOTAL BUSINESS ASSETS	1,219,225	1,346,264			
			Deferred tax on non-current assets		53,302
Personal Assets		21,900			
			TOTAL LIABILITIES	1,124,124	1,177,426
			Owner equity:		
			Retained Earnings	75,101	75,101
			Contributed Capital	20,000	20,000
			Personal Net Worth	–	21,900
			Valuation Equity	–	73,737
TOTAL ASSETS	1,219,225	1,368,164	TOTAL OWNER EQUITY	95,101	190,738
			TOTAL LIABILITIES & OWNER EQUITY	1,219,225	1,368,164

Figure 3.5. December 31, 19X2, projected balance sheet, Snake Belly Farms.

profit, Joe will plant no corn in 19X2 and plans to buy corn in January 19X3 for his steers. Joe sold all of his cotton and cottonseed at harvest and all of his wheat within two months of harvest. Stocker steers were purchased in October to replace the steers sold in June.

Joe projects that if the price and yield projections occur, he will end 19X2 with a cash balance of $13,300, net farm income of $29,727, and market value owner equity of $190,738. His ending financial position is somewhat stronger than his beginning position as evidenced by the financial ratios and comparisons presented in Table 3.4.

STATEMENT OF CASH FLOWS
Snake Belly Farms
Projected 19X2

CASH FLOWS FROM OPERATING ACTIVITIES:			
Cash received from farm operations:			
Feeder livestock and poultry sales	247,752		
Crops and feed	242,717		
Government payments, cash and certificates	23,392	513,861	
Cash received from non-farm income and operations:			
Wages		14,400	
Cash paid for farm operating activities:			
Feeder livestock and poultry	173,250		
Feed purchases	4,594		
Interest expense	106,189		
Operating expenses	162,785	-446,818	
Income and Social Security Taxes		-1,000	
NET CASH INCOME			80,443
Cash withdrawals for family living		-18,000	
NET CASH PROVIDED BY OPERATING ACTIVITIES			62,443
CASH FLOWS FROM INVESTING ACTIVITIES:			
Cash paid to purchase:			
Machinery and equipment		17,000	
Farm real estate: other farm assets		3,710	
NET CASH PROVIDED BY INVESTING ACTIVITIES			-20,710
CASH FLOWS FROM FINANCING ACTIVITIES			
Operating and CCC loans received (including interest paid by loan renewal)		87,576	
Term debt financing-loans received		168,000	
Operating debt principal payments		-87,576	
Term debt principal payments: Scheduled payments		-206,433	
NET CASH PROVIDED BY FINANCING ACTIVITIES			-38,433
NET INCREASE (DECREASE) IN CASH AND CASH EQUIVALENTS			3,300
Cash and cash equivalents reported on the beginning-of-year balance sheet:			10,000
Cash and cash equivalents, as calculated, at the end of year			13,300

Figure 3.6. 19X2 projected statement of cash flows, Snake Belly Farms.

SUMMARY

The Snake Belly Farms case study provides a comprehensive example of a forward plan that will be developed in detail throughout the remaining chapters of this book. Joe Farmer followed the twelve-step planning process outlined in Chapter 2, and at the risk of oversimplification, all we have to do now is to see how Joe put all of the pieces together. In other words, we will take this completed plan and dissect and analyze the procedures used within each component of the planning process to justify the final results presented in this chapter.

Table 3.4. Financial ratio analysis of Snake Belly Farms, December 31, 19X2, with comparisons

	Value 12-31-X1	Value 19X2	Value 12-31-X2	Percentage Change
Liquidity Ratios				
1. Current Ratio	1.15		1.20	+4.3%
2. Working Capital	34,203		43,032	+25.8%
Solvency Ratios				
3. Debt/Asset Ratio	0.89		0.86	- 3.6%
4. Equity/Asset Ratio	0.11		0.14	+30.2%
5. Debt/Equity Ratio	8.34		6.17	+26.0%
Profitability Ratios				
6. Rate of Return on Farm Capital		8.64%		
7. Rate of Return on Farm Equity		8.07%		
8. Operating Profit Margin Ratio		0.23		
9. Net Farm Income		29,727		
Repayment Capacity				
10. Term Debt and Capital Lease Coverage Ratio		0.55		
11. Capital Replacement and Term Debt Repayment Margin		18,091		
Financial Efficiency				
12. Asset Turnover Ratio		0.37		
13. Operational Ratios				
a. Operating Expense Ratio		0.38		
b. Depreciation Expensse Ratio		0.08		
c. Interest Expense Ratio		0.21		
d. Net Farm Income from Operations Ratio		0.06		

NOTES

1. Refer to a financial analysis text for a further description of ratio analysis. Suggested readings include J.D. Libbin and L.B. Catlett, *Farm and Ranch Financial Records,* Macmillan, New York, N.Y., 1987, Chapter 22; and J.D. Penson, Jr. and C.J. Nixon, *How to Analyze Financial Statements in Agriculture,* Agri-Information Corporation, College Station, Tex., 1985.

2. The ratios used in Table 3.1 are those recommended by the Farm Financial Standards Task Force.

PART II

Plan Components

The first three chapters of this book have outlined the purposes, processes, and structure of the cash flow statement as well as the ties between the cash flow statement and the other major financial statements. Despite all of this discussion, we still feel the bewilderment expressed by William Loscalzo[1]—the difficulty in making the transition from the mechanics of cash flow to preparing a forecast, essentially from scratch.

Each of the next five chapters will concentrate on one aspect of preparing the cash flow plan by expanding the components of the plan presented in the case study of Chapter 3. The purpose of each chapter will be to expose the myriad assumptions made in Chapter 3 and to discuss how to predict, project, or find the information needed to complete a comprehensive cash flow plan for a specific business. One important component, financial planning, will be delayed to Part III, Chapter 9, mainly because we must put all of the other components together before we begin the financing process and because the financing plans are one of the major tools for adjusting cash inflows to cover deficits and wisely use surpluses.

NOTE

1. See Preface, p. ix.

RECOMMENDED READINGS

Ross O. Love, Michael L. Hardin, and Harry P. Mapp, Jr., *Developing a Cash Flow Plan*, OSU Extension Facts No. 751, Cooperative Extension Service, Oklahoma State University, Stillwater, Okla., July 1983.

Mark Wilsdorf, "Whole Farm Cash Planning," *AgriComp*, p. 24.

John B. Penson, Jr., and Clair J. Nixon, *How to Analyze Financial Statements in Agriculture*, Agri-Information Corporation, College Station, Tex., 1985.

A. Gene Nelson and Thomas L. Frey, *You and Your Cash Flow*, Century Communications, Inc., Skokie, Ill., 1983.

Timothy G. Baker and John R. Brake, *Cash Flow Analysis of the Farm Business*, Extension Bulletin E-911, Cooperative Extension Service, Michigan State University, East Lansing, Mich., October 1975.

Billy B. Rice, "Cash Flow Planning: A Worksheet," *Farm Management Planning Guide*, Section YIII, No. 1. Cooperative Extension Service, North Dakota State University, Fargo, N.Dak., March 1982.

Estel H. Hudson and Clark D. Garland, *Cash Flow Analysis: A Management and Borrowing Tool*, Publication 708, University of Tennessee, Knoxville, Tenn., 1975.

Richard W. Wilt, Jr., and Sidney C. Bell, *Use of Cash Flow Statements as a Financial Management Tool*, Bulletin 487, Auburn University, Auburn, Ala., 1977.

CHAPTER 4

Strategic Planning

Many, if not virtually all, of the major corporations in the world today make time for the upper echelon of their corporate managers to leave day-to-day pressures and normal routines behind and remove themselves to a non-office location, all in an effort to allow them the time to rethink where the company has been going and the paths taken. Once just an academic concept, this process, now known as strategic planning or strategic management, has developed into several widely read journals, books, college-level courses, and continuing education programs and seminars, as well as into a widely practiced discipline in industry. Massive amounts of evidence incontrovertibly indicate that business leaders worldwide believe that the time taken away from the normal routine for strategic planning is the most worthwhile, valuable time spent each year. Despite this huge body of evidence as to the time and effort spent by corporate leaders in strategic planning, few owner-operated businesses advocate or practice the strategic planning process. Neither do virtually any farmers and ranchers, other than the industry leaders, of course. In just the past two or three years, farm and ranch executive programs have come alive.[1] All spend significant time helping participants see the value of strategic planning as well as teaching them how to begin.

In a more colloquial way, we used to summarize the strategic planning process into, and call it, goal setting. The discipline of strategic management actually goes far beyond what we called goal setting, but regardless of the name, the concept is the same. Business owners, executives, farm and ranch business owner-operators, corporate officers, farm and ranch partners, family members with an ownership or operation stake in the business, and all other management personnel need to take some time out to study, think, plan, and organize their businesses. This text cannot duplicate all of the strategic planning literature, but we will summarize the importance of the concept as we turn to the need to set goals.

GOAL SETTING

Good managers know the importance of a plan. They know that the components of a plan are simply goals, procedures, and a time frame for completion. A goal is *what* you want to do, a procedure is *how* you are going to achieve the goal, and the time frame for completion is *when* the goal will be achieved.

A goal can also be thought of as a destination—*where* you are going. Whether you view goals as what or where is immaterial. What is important is that you identify the end result—this is what I am going to do or this is where I am going. Gessaman and Prochaska-Cue point this out very clearly with the following example:

> Imagine that you are on an airplane that has just taken off from Honolulu for a non-stop flight to the mainland. What would you think if this announcement came over the intercom:
>
>> "Ladies and gentlemen, this is your captain speaking. We're now traveling east across the Pacific Ocean. If all goes well, we soon should be able to look down and see land below us. If we do, we'll hunt for a city that has an airport. When we find one that we think looks good, we'll land and find out where we are. Then, we'll decide where we want to go on the next leg of this flight. Meanwhile, folks, just sit back, relax, and enjoy the flight. The jet stream is behind us today, and we'll soon be traveling more than 550 miles per hour."
>
> Would you want to ride on that airplane? How confident would you be that you would arrive safely and on-time? Would you wonder why it was an advantage to travel at 550 miles per hour when you weren't sure where you were going?[2]

This example is a powerful reminder of the necessity of goals—yet we often neglect goals as unimportant in our personal and business lives.

We set goals consciously and unconsciously. Unconscious goal setting is often thought of as being unproductive, and by some, to not exist. They argue that if you do not set your goals consciously, then you really are not setting goals. Consider for example someone who says that they are going shopping. Shopping is an action, that is, a procedure to do something—not a goal. Although some would argue that the person's goal is to go shopping, goals are generally not actions. When pressed further, the person going shopping probably has an unconscious goal of "I want to relax," or "I want to people watch," or "I want to be active." Thus going shopping is *how* they reach their goal.

Unconscious goals such as these are the most common type. The key is to think about them and move them from the subconscious to the conscious. Once we have begun to think about our goals, the next step is to make sure they are meaningful.

Be Specific

If a farm or ranch operator said his or her goal was to make more money, the next obvious question is *how*? Plant more acres? Change to different crops? Get a job in town? Put the children to work? Having a non-specific goal doesn't help much. It makes the *how* too broad and thus difficult to achieve. Learn to set meaningful, specific goals such as: I want to increase my net farm income $10,000. Once a specific goal has been set, then the other two critical components of the plan can be added.

GOAL *WHAT:* Increase net farm income $10,000

PLAN *HOW:* Enhance marketing skills and sell all commodities in top one-third
of yearly price distribution
WHEN: December 31 this year

Personal versus Business

Personal goals can be the same as business goals; however, it will prove more helpful in the long run if they can be separated. This is perhaps the most difficult part of goal setting. You must separate your personal goals from your business, at least initially. It may prove necessary to later blend the two into one overall set of goals, but at the beginning of the planning process the two need to be separated. Figure 4.1 shows that personal and business goal plans can be totally different, partially the same, or totally the same.

Consider the previous example: the goal of the farmer was to increase net farm income $10,000. This was the business's goal—or was it? Joe and Mary Farmer own and operate the farm. What were their personal goals? When pressed to develop personal goals, Joe said what he really wanted was to buy a new pickup truck with the extra money earned this year. He was tired of driving a 15-year-old truck into town. All of his high school friends saw the old pickup, and Joe felt they thought he wasn't successful. Mary wanted to attend a CPA Exam Review shortcourse in San Diego (she works for a local accounting firm). Both Joe and Mary wanted the business to make more money for *personal* reasons. This is neither wrong nor inconsistent considering that the farm is a sole proprietorship. But because they separated personal goals from business goals, an obvious conflict arises. For both Joe and Mary to achieve their personal goals, the business must make more than $10,000—that is, $10,000 will not buy both a new pickup *and* send Mary on her trip. Thus they must compromise their personal goals and/or change their business goals.

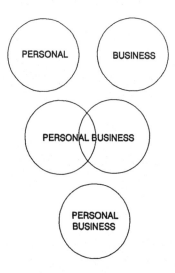

Figure 4.1. Personal and business goal plans.

It is important to separate personal and business goals even if the business structure is a partnership or corporation. Partnerships are particularly troublesome because they involve two or more different individuals and thus two or more attitudes about the business and two or more sets of personal/family goals. This is also true for corporations, especially family corporations. Because the process is more difficult does not mean it shouldn't be done; on the contrary, it means it's all the more important.

Short-Run versus Long-Run Goals

In addition to setting meaningful and specific personal and business goals for the short run, it is also advisable to set longer-term goals. Usually the most helpful planning horizon is approximately five years. If you plan much longer than five years it becomes very difficult to visualize what possibly might happen. Five years is long enough to force you into proper sequential thinking and is short enough that you feel you aren't just wastefully dreaming.

Our example farm's current short-run goal is to increase net farm income $10,000 (subject to potential revision after Joe and Mary reevaluate their personal goals). Where do Joe and Mary want their farm to be in five years? Just as importantly, where do Joe and Mary want to be personally in five years? Remember that it's just as important to set meaningful and specific long-term goals as it is short-term goals. Joe and Mary's long-term goal for their farm may be simply to still be in business in five years. This truly may be their long-term goal, but it isn't specific. Thus *how* to still be in business in five years becomes a difficult question to answer. Perhaps a more meaningful specific long-term goal might be: Snake Belly Farms will have net farm income of $30,000.

FIVE-YEAR GOAL PLAN   *WHAT*:   Have net farm income of $30,000
                       HOW: Diversify into high-value fruit or vegetable crops
                       WHEN: December 31, in five years

This is a specific goal and, if successfully attained, should meet the general goal of still being in business in five years.

Is the long-term business goal consistent with Joe and Mary's personal long-term goals? Joe's long-term goals are to be a successful marketer and to drive a new pickup at least every three years to maintain his personal image in the community. Mary wants to have her CPA designation, have a personal computer to do the business records, and to do accounting and record-keeping consulting for other farmers and ranchers part time. Are the personal and business goals compatible? Perhaps, if the $30,000 net farm income each year will provide the things that they want personally.

THE GOAL-SETTING PROCESS

Goal setting involves *WHAT, HOW,* and *WHEN*: setting meaningful, specific goals; separating personal and business goals; and setting short- and long-term goals. The goal-setting process is merely recognizing that all of the above components do not stand alone and must be considered while viewing the other parts. Figure 4.2 shows the interaction of the goal-setting process. Notice that the process proceeds from one goal *towards* the other goal. This is important because it places the goal where it belongs in the hierarchy. That is, you should evaluate whether a personal goal is possible by looking at the business's goal *and then* look at the business's procedures and timing. Don't jump from the personal goal to the business's procedure, because you have then bypassed what is most important: the *goal*. This keeps the process goal oriented.

Finally, the process should evaluate the short-term goal(s) relative to the longer-term goal(s), again looking first at the goals of each, not the procedures and timing. Goals are the central focus. They often can be achieved several ways. The old saying, "There is more than one way to skin a cat" is very true in most cases. There are usually several possible procedures and many time frames to meet most meaningful goals.

Figure 4.3 presents a generic form to help the goal-setting process along. A form like this should be reproduced and used for both personal and business and for short-run and five-year plans. Simply circle whether it is personal or business and either short-run or for five years. The process is most valuable if done separately; that is, develop the personal goal plan and then the business goal plan (or vice versa). For a family, it is critical that the initial goal plans be developed individually. One or both parties may not be totally willing to express their true attitudes if the initial process involves working together. However, they *must* work together once they have individually determined personal and business goal plans.

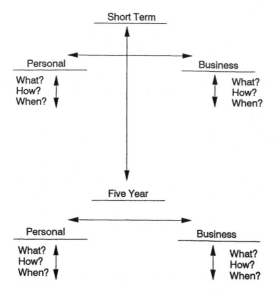

Figure 4.2. The goal-setting process.

PROCESS EXAMPLE: SNAKE BELLY FARMS

Once each individual has developed personal and business goal plans, then the process can begin to compromise, consolidate, rethink procedures/timing, and to even start the whole process over again. Figures 4.4 and 4.5 show Joe Farmer's personal goal plans for both short-term and five-year time periods, and Figures 4.6 and 4.7 repeat the process for Mary. Figures 4.8 and 4.9 show Joe and Mary's short-term and five-year business goal plans.

Let's look at Joe's personal goal plans first (Figure 4.4). Joe has a very specific short-term goal—to buy a new pickup. He is going to pay for it from the profits the farm earns (so he thinks), and he is going to get the pickup in the fall. However, Joe's five-year personal plan isn't as specific except for the purchase of a new pickup every three years. Joe's long-run personal goal is to sell all of his commodities in the top one-third of the annual price distribution. He has a general goal and some specific procedures. Joe needs to be more specific with his overall goal so that it is easier to achieve. Perhaps a more specific, meaningful goal for Joe would be to:

1. Increase knowledge of forward contracts and seasonal price distributions.
2. Increase knowledge of hedging crops and livestock with futures and options.
3. Actively use seasonal price movements, price cycles, and hedging with futures and options to market all farm products produced.
4. Buy a new pickup every three years.

Notice that Joe's general goal of being a better marketer hasn't necessarily changed, but it has been restated in terms that are measurable. Joe can then compare this performance with county statistics to see if he really is a better marketer.

Mary, on the other hand, has fairly specific, measurable long-term goals but a general difficult-to-measure short-term goal—to keep her accounting skills sharp. Mary needs a more measurable one-year goal such as:

1. Prepare local farmers' tax returns
2. Help local farmers set up record-keeping systems

These goals would certainly help Mary keep her accounting skills sharp but may or may not require her to go to San Diego for a week-long course. However, in looking at Mary's five-year goals, she wants to pass the CPA exam, and for her to pass it this year requires the course in San Diego.

Joe and Mary's business goals are now specific and measurable. Are they consistent with their personal plans? Maybe—if they achieve the $30,000 net farm income, they may be able to do what both want to do (Mary's plan calls for about $7,400 and Joe's about $2,200 plus the purchase of two pickups), depending upon living expenses. Their short-term plans are certainly not financially possible unless they finance Joe's pickup with a bank loan. Thus Joe and Mary now need to sit down again and reconcile their short-term personal goals and their business one-year goal plan. They need to see if the business plan is compatible with their five-year personal plans both financially and in terms of timing. For example, Mary wants to start her own financial records consulting business at the same time that Joe is taking extension courses—is there enough time for them to do both? Is there enough money? The

Underline which plan (short-term or five-year plan) and whether it is for personal or business goals.

1. SHORT-TERM (PERIOD FROM   TO   )

    FIVE-YEAR

2. WHAT DO YOU WANT OR WHERE DO YOU WANT (YOUR BUSINESS) (YOU PERSONALLY) TO BE AT THE END OF THIS PERIOD?

3. HOW DO YOU PLAN TO ACHIEVE THE ABOVE GOAL(S)?

4. WHEN? WHAT IS THE TIME TABLE FOR EACH PROCEDURE?

Figure 4.3. Goal plan for developing personal and business goals for short-term and five-year plans.

Underline which plan (short-term or five-year plan) and whether it is for personal or business goals.

1. SHORT-TERM (PERIOD FROM January 1 TO December 31)

    FIVE-YEAR

2. WHAT DO YOU WANT OR WHERE DO YOU WANT (YOUR BUSINESS) (YOU PERSONALLY) TO BE AT THE END OF THIS PERIOD?

    Buy new pickup

3. HOW DO YOU PLAN TO ACHIEVE THE ABOVE GOAL(S)?

    From profits from farm

4. WHEN? WHAT IS THE TIME TABLE FOR EACH PROCEDURE?

    When new models come out, fall, current year

Figure 4.4. Joe Farmer's short-term personal goals.

Underline which plan (short-term or five-year plan) and whether it is for personal or business goals.

1. SHORT-TERM (PERIOD FROM   TO   )

   <u>FIVE-YEAR</u>

2. WHAT DO YOU WANT OR WHERE DO YOU WANT (YOUR BUSINESS) (<u>YOU PERSONALLY</u>) TO BE AT THE END OF THIS PERIOD?

   1. Sell all commodities in top one third of annual price distribution
   2. Have a new pickup every three years

3. HOW DO YOU PLAN TO ACHIEVE THE ABOVE GOAL(S)?

   1. Take university extension courses--series of five offered every two months (two days each)--approximate cost $800
   2. Subscribe to newsletters, magazines, etc. ($200)
   3. Attend state and national meetings ($1200)
   4. Trade old pickup in and finance the balance from profits of the farm

4. WHEN? WHAT IS THE TIME TABLE FOR EACH PROCEDURE?

   1. Courses--complete by fall, second year
      Attend state convention each year
      Attend national convention every other year

   2. Get new pickup current year and trade it in fall fourth year

Figure 4.5. Joe Farmer's five-year personal goals.

Underline which plan (short-term or five-year plan) and whether it is for personal or business goals.

1. <u>SHORT-TERM (PERIOD FROM January 1 TO December 31)</u>

   FIVE-YEAR

2. WHAT DO YOU WANT OR WHERE DO YOU WANT (YOUR BUSINESS) (<u>YOU PERSONALLY</u>) TO BE AT THE END OF THIS PERIOD?

   Keep accounting skills sharp

3. HOW DO YOU PLAN TO ACHIEVE THE ABOVE GOAL(S)?

   Attend a week long workshop in San Diego

4. WHEN? WHAT IS THE TIME TABLE FOR EACH PROCEDURE?

   Summer of this year

Figure 4.6. Mary Farmer's short-term personal goals.

Underline which plan (short-term or five-year plan) and whether it is for personal or business goals.

1.   SHORT-TERM (PERIOD FROM   TO   )

FIVE-YEAR

2.   WHAT DO YOU WANT OR WHERE DO YOU WANT (YOUR BUSINESS)
      (YOU PERSONALLY) TO BE AT THE END OF THIS PERIOD?

    1.   Get CPA certification
    2.   Get personal computer with accounting software
    3.   Computerize business records of farm
    4.   Become a consultant to other farms and ranches on computer record system

3.   HOW DO YOU PLAN TO ACHIEVE THE ABOVE GOAL(S)?

    1.   Attend week-long seminar in San Diego, take CPA exam, pass ($1,200)
    2.   Review software and hardware, purchase system ($4,000)
    3.   Work nights/weekends to computerize from records ($2,000)
    4.   Get business licenses, advertise in local paper and word of mouth ($2,000)
    5.   When business is large enough, quit part-time job to run the consulting business
        Funds for above estimated to be approximately $7,400; from profits of farm

4.   WHEN? WHAT IS THE TIME TABLE FOR EACH PROCEDURE?

    1.   Summer, current year, for seminar. Pass CPA by fall, current year
    2.   By December 31, second year
    3.   By summer, third year
    4.   By fall fourth year, have consulting business in operation and be self-supporting
        by summer, fifth year

Figure 4.7. Mary Farmer's five-year personal goals.

Underline which plan (short-term or five-year plan) and whether it is for personal or business goals.

1.   SHORT-TERM (PERIOD FROM January 1 TO December 31)

FIVE-YEAR

2.   WHAT DO YOU WANT OR WHERE DO YOU WANT (YOUR BUSINESS)
      (YOU PERSONALLY) TO BE AT THE END OF THIS PERIOD?

Net farm income increase of $5,000

3.   HOW DO YOU PLAN TO ACHIEVE THE ABOVE GOAL(S)?

Crop enterprise budgets indicate that the expansion will yield a net income of $5,620

4.   WHEN? WHAT IS THE TIME TABLE FOR EACH PROCEDURE?

Have new crop mix in place by March 15

Figure 4.8. Short-term business goals, Snake Belly Farms.

Underline which plan (short-term or five-year plan) and whether it is for personal or business goals.

1.   <u>SHORT-TERM (PERIOD FROM January 1 TO December 31)</u>

     FIVE-YEAR

2.   WHAT DO YOU WANT OR WHERE DO YOU WANT <u>(YOUR BUSINESS)</u> (YOU PERSONALLY) TO BE AT THE END OF THIS PERIOD?

     Have net farm income of $30,000 (currently $12,000)

3.   HOW DO YOU PLAN TO ACHIEVE THE ABOVE GOAL(S)?

     1.   Decrease production of wheat by 50% and add acreage in one or more of the folowing: carrots, lettuce, and onions
     2.   Decrease hay production by at least 50% and replace with either corn or cotton
     3.   Expand stocker-feeder operation through leased pasture when feasibility is positive
     4.   Enhance financial planning and management expertise in order to conserve financial resources and invest wisely

4.   WHEN? WHAT IS THE TIME TABLE FOR EACH PROCEDURE?

Figure 4.9. Five-year business goals, Snake Belly Farms.

process continues until personal and business goal plans are compatible. This certainly will involve compromise, sacrifice, and rethinking.

FLEXIBILITY

Goal setting minimizes surprises and introduces flexibility into both the personal and business aspects of Joe and Mary's life. Can you imagine the conflict that could occur this summer and fall when Mary decides to go to the San Diego conference costing approximately $1,200 and Joe (who is very sensitive about his 15-year-old pickup) buys a new truck? The goal-setting process won't eliminate conflicts, but it does help minimize them. And if both Joe and Mary have done an honest job of goal planning, then they will have done a lot of thinking about priorities, different ways to do things, and timing. Thus when a problem occurs, when a goal can't be met, they are in a much better position to rethink things, reestablish new goals and/or procedures or timing, and move forward. The problems arise when people *don't* think about them and thus don't develop goal plans.

SUMMARY

Goal setting is an extremely important part of the business planning process. The goal-setting process must determine both personal and business goals of all the principal individuals involved in the business. The goals must be specific and should include both short- and long-term goals and must include a set of procedures and a timetable for *how* and *when* the goals are to be achieved. Each individual should determine his or her goals separately and then meet to discuss any conflicts of time or resources, to set joint goals, and to find ways to meet as many goals as possible. Finally, all goals should be flexible so that plans can be adjusted to changing conditions or priorities.

NOTES

1. An excellent example of a recently developed program is the Texas A&M University-sponsored Executive Program for Agricultural Producers, organized by D.A. Klinefelter.

2. P.H. Gessaman and K. Prochaska-Cue, *Goals for Family and Business Financial Management,* Cooperative Extension Service Report CC312, University of Nebraska-Lincoln, rev., July 1985.

RECOMMENDED READINGS

Paul H. Gessaman and Kathy Prochaska-Cue, *Goals for Family and Business Financial Management,* CES Report CC312, Cooperative Extension Service, University of Nebraska-Lincoln, rev., July 1985.

R. Patrick Sullivan and James D. Libbin, *Goals of New Mexico Farmers,* Research Report 606, Agricultural Experiment Station, New Mexico State University, Las Cruces, N.Mex., May 1987.

A. Gene Nelson, *Setting Farm Business Goals,* Extension Circular 1097, Cooperative Extension Service, Oregon State University, Corvallis, Oreg., November 1981.

Basudeb, Biswas, John R. Lacey, John P. Workman, and Francis H. Siddoway, "Profit Maximization as a Management Goal on Southeastern Montana Ranches," *Western Journal of Agricultural Economics,* 9:1, July 1984, pp. 186-194.

Wyatte L. Harman, Roy E. Hatch, Vernon R. Eidman, and P.L. Claypool, *An Evaluation of Factors Affecting the Hierarchy of Goals,* Oklahoma Agricultural Experiment Station Technical Bulletin T-1347, Stillwater, Okla., June 1972.

Wilmer M. Harper, and Clyde Eastman, "An Evaluation of Goal Hierarchies for Small Farm Operators," *American Journal of Agricultural Economics,* 62:4, November 1980, pp. 742-747.

James B. Kliebenstein and George F. Patrick, *Multiple Goals and Attitudes of Farm Decision Maker,* Ag Econ Paper No. 1980-25, An Executive Summary, University of Missouri-Columbia, 1980.

J.O. Wise and R.L. Brannen, "The Relationship of Farmer Goals and other Factors to Credit Use," *Southern Journal of Agricultural Economics,* 15:2, December 1983, pp. 49-54.

W. Graham Astley et al., "Complexity and Cleavage: Dual Explanations of Strategic Decision-Making," *Journal of Management Studies,* 19:4, October 1982, pp. 357-375.

Matt Gibbs and Toby Kashefi, "A Cause of Business Failures," *Pittsburgh State University Business and Economics Review,* 8:2, November 1982, pp. 3-11.

Frederick Gluck, Stephen Kaufman, and A. Steven Walleck, "The Four Phases of Strategic Management," *Journal of Business Strategy,* 2:3, Winter 1982, pp. 9-21.

William F. Glueck, *Business Policy and Strategic Management,* McGraw-Hill, New York, N.Y., 1980.

James M. Higgins, *Organizational Policy and Strategic Management: Text and Cases,* Dryden Press, Chicago, Ill., 1983.

LaRue Tone Hosmer, "The Importance of Strategic Leadership," *Journal of Business Strategy,* 3:2, Fall 1982, pp. 47-57.

Milton Leontiades, "A Diagnostic Framework for Planning," *Strategic Management Journal,* 4:1, January/March 1983, pp. 11-26.

Albert O. Trostel and Mary Lippitt Nichols, "Privately-Held and Publicly-Held Companies: A Comparison of Strategic Choices and Management Processes," *Academy of Management Journal,* 25:1, March 1982, pp. 47-62.

CHAPTER 5

Market Planning

Someone once said that forecasting is dangerous, especially about the future. Yet we must forecast and engage in forward analysis. Forward analysis is nothing more than looking at possible futures—what might happen and the implications of the possible or likely outcomes. If we don't actively try to project the future then we do it by default. For example, if you don't actively try to predict what corn prices will be next fall, then by default you have made a prediction, i.e., that they will be the same as now or the same as last season. No forecast or forward analysis will always be perfect, but it doesn't have to be perfect, or even correct, to be valuable. Good market planning rests firmly on good forecasting and forward analysis. One of the principal reasons why most individuals and firms are very poor at marketing is because they resist forward analysis.

FUNDAMENTALS OF A MARKETING PLAN

Market planning is a form of forward analysis that involves three simple steps: 1) situation analysis; 2) establishment of goals; and 3) development of marketing strategies.

Situation Analysis

A situation analysis is best begun with a series of questions: What do I have to work with? How have I done it in the past? What are my alternatives in the future? Each of these questions can be further broken down:

1. What do I have to work with?
   a. Inventory of physical assets
   b. Inventory of management/marketing skills

2. How have I done it in the past?
   a. Inventory of currently produced inputs
   b. Inventory of currently purchased inputs
   c. Inventory of current sales procedures
   d. Inventory of current purchasing procedures
3. What are my alternatives in the future?
   a. Inventory of potential new enterprises
   b. Market channel analysis of current and potential enterprises
   c. Evaluation of the pricing tools available for each enterprise

Situation analysis is one of the most difficult parts of market planning because it requires the combination of a great amount of information along with critical thinking. Oftentimes when we have completed a good situation analysis we feel that we have done all of the marketing we need to do—we know what we have done in the past and what potential alternatives exist for the future. However the situation analysis cannot stand alone—it only lists potential alternatives—it doesn't formulate a marketing goal plan. The more time spent doing a situation analysis, the easier it is to formulate marketing goals and plans of action.

Goals

As discussed in Chapter 4, a goal is simply what you want to do or where you want to go. It is not uncommon for producers to believe that they have a marketing goal. Often that goal is to receive a better price, or even more unrealistically, to top the market. Both of these are common goals yet they should more properly be called wishful thinking or dreaming. Neither are meaningful or specific. A good, detailed situation analysis is the major starting point to develop marketing goals.

Consider Joe Farmer, who has completed a detailed situation analysis for his farm. In looking at his alfalfa enterprise he finds that he received an average price of $96 per ton last season. Seasonal average price charts show that his price was below average. That is, the average price received was $100 per ton. Suppose that Joe establishes this year's market goal for alfalfa as receiving at least $100 per ton.

At first blush this appears to be a very good goal. Upon closer observation, however, it is not. It is specific—at least $100 per ton—but it is not meaningful. Market forces will change the price of alfalfa each year and thus some forecast needs to be made to make the goal reasonable. A more appropriate goal would be to receive the seasonal average price for alfalfa rather than the below-average price received last year. A reasonable forecast of what the seasonal average price of alfalfa will be next year is needed and that figure should be used in the market goal for alfalfa rather than the wishful $100 per ton. If we don't actively forecast alfalfa prices for next year then we have done so by default, i.e., we naively assume that next year's prices will be the same as last year's.

Marketing goals need to be established for each enterprise that the farm or ranch plans to produce. Once each goal has been established, then a marketing strategy can be developed.

MARKETING STRATEGIES

A market strategy is nothing more than a procedure or set of procedures to achieve a stated goal. A strategy differs from a procedure in that strategies involve "if, then" criteria. That is, if some event occurs, then follow procedure A; however, if it does not occur, then follow procedure B.

Suppose Joe establishes a marketing goal of receiving at least $3.20 per bushel for his corn. A possible corn marketing strategy might be:

1. If during the period January 1 to April 15, the December corn futures price reaches $3.45 plus or minus 3 cents, the entire projected corn crop (35,000 bushels) will be hedged. The hedge will be maintained until harvest and the crop sold at local market prices. Projected net price will be $3.20 plus or minus 3 cents using an estimated basis of 25 cents per bushel.

2. If during the period January 1 to April 15 the December corn futures price does not reach $3.45 plus or minus 3 cents, then the crop will remain unhedged with futures. A put option on December futures will be purchased no later than April 30 that yields the highest minimum price floor for a premium of no more than 5 cents per bushel. The crop will continue to remain unhedged with futures unless a pricing opportunity exists to hedge during the growing season at a price of $3.45 plus or minus 3 cents. The crop will be sold at harvest to local cash markets. Projected net price will be the minimum price established with the put option on December corn futures.

The marketing strategy for the corn marketing goal is fairly simple. The goal is to receive $3.20. The strategy is to hedge with futures when the opportunity exists to receive $3.20, but if market conditions are such that $3.20 is unattainable, then options will be used to get the highest minimum price obtainable. A marketing strategy like this one allows more flexibility and a better chance of meeting the marketing goal for corn. One procedure for the goal might have been listed as: Hedge at planting time with December corn futures. But if this was our only goal, then we might have hedged at a very low price relative to $3.20 and lost flexibility. Of course we could modify these goals and procedures once we see what prices are at harvest, but then the goal would essentially be meaningless. Market strategies, on the other hand, offer the manager the flexibility that if forecasts are wrong, an alternative has been thought about that is acceptable and offers the best possibility of achieving the stated market goal. When possible, therefore, develop market strategies, not single procedures.

MARKETING TOOLS

Let's look at some of the more important tools used in developing a marketing plan. The tools discussed are by no means all of the tools used but represent some of the more popular and useful tools.

Market Channels

A market channel is the system that a product goes through from production to final consumption. In the strict marketing sense it involves the inputs used in the production process and the institutions and functions involved in getting the product to the consumer. Thus a generalized market channel would be:

Input Suppliers → Producers → First Handlers →
Processors → Wholesalers → Retailers → Consumers

The participants in this market channel are organized into various institutional forms and perform numerous types of functions.

INSTITUTIONS

A producer must deal with input suppliers and first handlers of his product. What type of business form are these firms and in what type of market structure do they operate? Studying these institutional forms does not necessarily imply an attempt to change them, but rather to understand more fully the extent of competition, pricing, and market alternatives. Two major types of institutions will be discussed—noncooperatives and cooperatives.

NONCOOPERATIVES

A noncooperative is characterized by sole proprietorships, partnerships, and corporations. They account for most of the types of firms operating in most product market channels. Of course, each product channel is unique and needs to be studied in detail. However, at issue in every product channel is the level of competition and its corresponding effect on the price of the item.

In looking at the corn market channel, we generally examine the first handler of the product. This is usually a local elevator but could also be a processor such as a feedyard, feedmill, or food processor. What choices are available? Several? Only a few? Generally, the larger number of potential choices for producers, the more competitive the price will be, i.e., higher than it would be with only a few choices.

One of the principal problems producers face is understanding how many choices actually exist. Most market channels have far more choices than a casual observation can find. We tend to believe we have only Hobson's choice (i.e., no choice at all). We feel we have to trade with a certain firm because they are the only firm or because they are the best firm. But Hobson's choice is really a choice—we can trade with a firm *or* choose not to trade with them and trade with someone else. Opponents may argue that there usually is no alternative place to trade. Only in very few cases is this true. A good market channel analysis looks for all potential choices.

We may believe that a farmer only has one local elevator to which to sell corn, but have we looked at surrounding areas? Sometimes the cost of transportation is less than the additional revenue received and thus may be cost-effective. Are there other potential first handlers? Is there a good grain broker who could provide a service? The point is that we need to continuously look for alternatives. Usually changing market conditions cause different alternatives to exist, and sometimes one will be better than another—rarely is one firm consistently better than another. In short, the process of searching for alternatives is usually cost-effective.

Market channel research is very important for traditional products, but it is critical for new products. For example, malting barley might be a potential new crop. The returns may look great, it may do very well in the area—but is there a market channel for the product? Where are the first handlers? Is the first handler in Denver? How do you get it there? Is it cost-effective? Often these questions kill the thought of any new or different product. But more than a casual look needs to be taken. If the new crop is very promising, perhaps pooling by individuals can make it cost-effective. Or perhaps an education program for the local elevator or processor on how to handle the product may make it possible. Market channel research requires work, but it also produces positive results.

COOPERATIVES

A cooperative is a business run by and for the people who use the cooperative. Cooperatives generally are a form of vertical integration that combine dissimilar market functions. A corn producer who joins and uses a local grain elevator cooperative is vertically integrating. The producer is combining the marketing function of selling the grain produced and the function of selling larger volumes of commingled grain to different outlets. The profits (or losses) that result from the vertical integration will go to each producer based upon how much of the business each person generated.

Cooperatives are prevalent in the local grain business, in providing inputs, and in some processing and service industries such as credit, insurance, and bargaining. Cooperatives can be very valuable to producers but, like all businesses, have problems. The major advantages of cooperatives include the following:

1. Cooperatives provide producers the opportunity to vertically integrate, offering the potential for profits from the integrated functions.
2. They can provide services that otherwise would not be offered or not offered as conveniently by other business forms.
3. They offer the flexibility to allow producers to band together for the common good of all in marketing, buying inputs, and/or providing services.
4. Cooperatives are democratically controlled, usually one person–one vote.
5. They can be used to successfully counter market power of other institutions and add competition.

The major disadvantages of cooperatives include the following:

1. They are difficult to organize and require considerable efforts by members to ensure they operate properly.
2. Cooperatives, like all businesses, require capital outlays and operating capital. The coop must either borrow this capital and/or keep a portion of earnings. In the early years of a cooperative or when an established cooperative is expanding, large portions of the profits that are due members may be retained by the cooperative for capital purposes. Thus the benefits of vertical integration may not be available immediately and in fact may not accrue for several years.
3. Sometimes in an effort to provide services, cooperatives expand into businesses that are not profitable and benefit only a few of the members.

PRODUCT DIFFERENTIATION

Participants in the market channel provide several different functions—sales, advertising, and exchanges, for example. However, one of the most important functions that is common throughout the channel for most products is product differentiation.

When looking at input purchases, keep in mind that most successful firms try to use product differentiation as a tool to sell products. Product differentiation can result from real differences that exist between products, and it can also exist when consumers feel there is a real difference when in fact the differences are nonexistent. Make sure that the difference in price between a brand name product and a non-brand name product is warranted by differences in quality. Often times that is very difficult to judge. If it is very difficult to judge, then perhaps the difference doesn't justify the extra expense.

Wise input marketers know how to ignore fancy promotions, giveaways, and slick advertising. They judge price, quality, and quantity and make decisions based upon objective criteria rather than emotion. Few if any sharp input marketers are brand loyal unless the brand consistently meets objective criteria.

Almost all major agricultural inputs have a significant amount of product differentiation that needs to be carefully analyzed by each individual whether or not the price differences are justified by quality differences.

Seasonal Price Movements

A seasonal price movement is the pattern that prices generally follow throughout one production period (usually one year). Almost all products have seasonal patterns. A simplified view of seasonal prices shows that they fall into two major categories—demand-driven and supply-driven.

Demand-driven seasonal price movements correspond to major changes in the demand for the product. Supply does not usually change that much, so most of the change in the price for the product comes from changes in demand. Figure 5.1 shows the hypothetical seasonal price movement for automobiles. The supply of automobiles is more or less fairly constant; however, demand changes as people change their tastes and preferences. In the fall, when new models come out, people's tastes and preferences generally favor a new or different car. However, when the new model year is ending in late summer, the car, although new, is almost one year old. With the new models coming out, demand decreases as people adjust their tastes and preferences to reflect their attitudes.

Supply-driven seasonal price movements are a result of major changes in supply rather than demand. Many agricultural products have supply-driven seasonal prices. The theory of supply-driven prices says that the seasonal price movement should correspond to the cost of production and/or the cost of storage. Figure 5.2 shows a seasonal price chart for corn that illustrates the concept. Prices are lowest at harvest in the fall and then generally increase throughout the year to reflect the cost of storing the corn.

Seasonal price movements can be a very useful marketing tool. They can help producers understand the periods when the price offers the best opportunity to increase or decrease. For livestock producers, knowing the seasonal price movement is necessary so that managers will have a better knowledge base for deciding whether

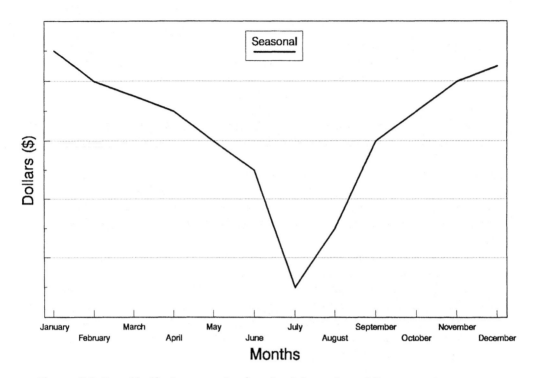

Figure 5.1. Hypothetical seasonal price chart for automobiles.

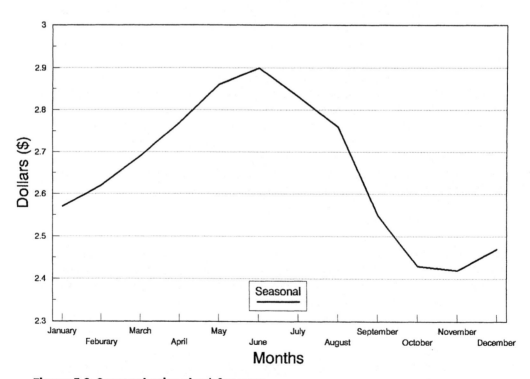

Figure 5.2. Seasonal price chart for corn.

or not to carry livestock to heavier weights, sell lighter, or take some other action. Seasonal price movements also show the tops and bottoms of markets and how long they generally last.

Cyclic Price Movements

A cycle is a repeatable price pattern that can be used to predict future price directions. The only known agricultural commodities that have cyclic price patterns are livestock. Efforts have been made to identify other agricultural commodities that have price cycles, but these commodities do not repeat the cycle with enough regularity to be useful as a predicting tool.

Figure 5.3 shows the cattle cycle (hogs and sheep also have cycles). The cattle cycle has been averaging between 9 and 12 years for the last four complete cycles. The price cycle is just the mirror image of the quantity cycle, i.e., as numbers are increasing, prices are generally decreasing, and vice versa. Cycles can be used like seasonal price trends to help estimate when to carry animals to heavier weights, to make weaning decisions, and to add or delete animal inventories.

Many successful producers try to operate in a counter cyclical fashion. They liquidate animal inventories when others increase inventories and vice versa. Thus they have animals for sale when prices are increasing. When prices are decreasing, they have adjusted inventories to be at their lowest so the number of animals for sale is less (or in some cases nonexistent). Of course, there are producers who ignore cycles and shoot for constant marketings year-round, year after year.

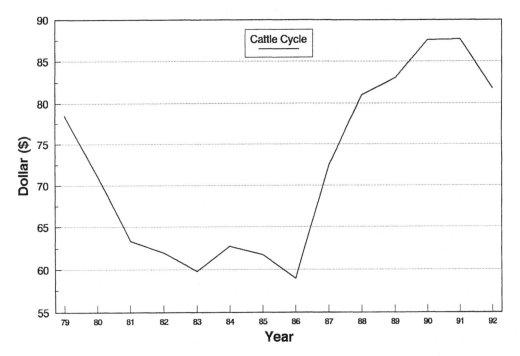

Figure 5.3. Cattle price cycle.

Delphi Approach

The Delphi Approach to forecasting is sometimes called the consensus or expert approach. It involves the collection of more than one forecast from individuals or firms. It is very similar to taking a consensus of race handicappers. Three of four of a certain race's handicappers may have picked the number four horse to win; thus the Delphi Approach says the best prediction for the winning horse is number four. However, if no consensus is possible, then no forecast is possible using the Delphi Approach.

The Delphi method is popular because it really doesn't require the user to forecast. The user is merely assembling forecasts from other sources and taking a consensus. A very handy way to use the Delphi method is to construct a table with the forecasts listed, as shown in Figure 5.4.

Fundamental Analysis

Fundamental analysis is simply the use of supply, demand, and other economic factors to predict prices. This usually involves a good knowledge of economics, statistics, and computer systems with results that usually don't justify the knowledge and work. However, a basic knowledge of fundamental analysis will let producers use some of the work done by others. In fact, we do not recommend that producers try to develop fundamental analysis skills; rather we advise them to concentrate on using the work that others (usually universities and consulting firms) have developed.

Simply stated, fundamental analysis involves expressing an economic relationship in the form of an equation and then using a statistical technique called regression to estimate the parameters of the equation using historical information. Once the parameters are estimated, they can then be used to forecast future values of the economic variables.

For example, the price of corn is believed to be inversely related to the quantity

FORECASTERS	FORECASTS		
	Direction	Percent Change	Time Frame
USDA	UP	+1	NEXT 6 MOS.
PRIVATE FORECASTER "A"	UP	+2	NEXT 12 MOS.
PRIVATE FORECASTER "B"	UP	+.05	NEXT 6 MOS.
PRIVATE FORECASTER "C"	DOWN	-.05	NEXT 6 MOS.

Figure 5.4. Delphi Approach matrix for corn price forecasts.

of corn on the market; that is, the more corn on the market, the lower the price. When we gather historical information, we find the following hypothetical results:

	Quantity Produced (billions of bu.)	Price (per bu.)
1980	8.0	$2.00
1981	6.9	3.20
1982	7.2	2.95
1983	8.2	1.95
1984	7.0	3.10
1985	6.5	3.45
1986	7.0	3.05

Thus our economic relationship is $P = f(Q)$, where P = Price of corn per bushel, Q = Quantity of corn sold, f = Functional form of the equation.

We can specify the functional form of the equation any way we want, but let's consider for the moment that it is linear and of the form $P = a + b(Q)$, where a is intercept term and b is the slope of the line. Thus our economic relationship is now expressed as $P = a + b(Q)$.

We can now use a statistical technique called regression to estimate the values of a and b, which might give us the following expression $P = 2.3 - .05(Q) + e$, where e = error term of regression. The error term is the catch-all expression, because regression gives us the best estimate of the relationship but not a perfect estimate. If the estimate were perfect, then there would be no error term. Now we can use the simple expression to forecast. We can get an estimate of the quantity of corn that will be sold, plug it into the equation, and get an estimate of the price. If the U.S. Department of Agriculture's estimate of corn production on April 1 for the coming harvest is 8.0 billion bushels, then $P = 2.3 - .05(8.0)$ yields an estimated price of corn at harvest of $2.33. Obviously, this is a very simplified forecast of corn prices. Other factors can and do have an impact on corn prices such as the price of other grains, livestock prices, and target prices. Thus the price equation for corn has other variables and becomes more difficult to estimate and use properly. However, the concepts remain the same. Table 5.1 shows the results of a model developed to forecast corn prices. Notice that it includes more than one variable, but the basic concepts are still simple. Most universities have price forecasting models that are available for use by producers. Also, most land-grant universities have extension personnel who will run the models for you and give you the forecasts. In addition, the USDA uses forecasting models to provide outlook information to producers.

For most producers, the time spent trying to forecast prices using fundamental analysis is better spent on other things—such as learning to interpret the results of fundamental analysis.

Technical Analysis

Because fundamental analysis requires so much time and effort, an alternative price forecasting method called technical analysis was developed. Technical analysis says that past price movements are an indication of future price direction, i.e., the past predicts the future. Technical analysis is also called charting analysis because most technical forecasters use charts to record price movements. Technical analysis

Table 5.1. Fundamental analysis used to predict corn prices

Estimated Coefficient	Variable
4.153	Intercept
-0.701	Corn loan rate (dollars per bushel) Source: *Agricultural Statistics*
-0.025	Estimated acres of corn planted in the United States (hundred thousand acres) Source: *Prospective Plantings*
+0.372	On- and off-farm grain stock inventory at the beginning of January (millions of bushels) Source: *Grainstocks* and USDA *Statistical Bulletin*
+0.412	January cash price of corn (dollars per bushel) Source: *Agricultural Statistics*
+0.911	0 if before 1973 and 1 if year is 1973 or after
-0.00004	Number of acres planted in corn lagged one year (thousand acres) Source: *Agricultural Statistics*
+0.0004	Slope shifter with a value of 0 from 1963 to 1971 and the volume of exports lagged one year from 1972 to 1984 (thousand bushels) Source: *Agricultural Statistics*

Source: M. Blake, J. Witte, and D. Truby, *Forecasting United States Feed Grain Prices,* Research Report 594, Agriculture Experiment Station, New Mexico State University, Las Cruces, N. Mex., 1986.

is most often used with futures prices, but it can also be used with cash prices.

Technical analysis usually involves bar charts, point and figure charts, and moving averages. In addition to these three types of technical analysis, there are many more types such as Elliot Wave Analysis, Fibonacci Sequence Analysis, and Relative Strength Index Analysis.

BAR CHARTS

A bar chart is a record of the high price during the day or week or month, the low price, and where the market settled. Figure 5.5 shows corn futures prices recorded on bar charts. Using the chart, trend lines can be drawn connecting the lows (for an up trend) or highs (for a down trend). When the trend line is broken, a reversal trend is forecasted. In addition to trend lines, other formations are used to predict price movements. They include gaps, island reversals, key reversals, hook reversals, head and shoulder formations, flags, pennants, and several more. All of these formations help the chartist predict future price directions based upon where the prices have been.

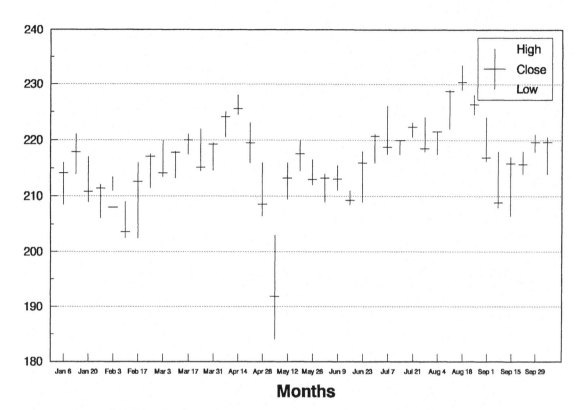

Figure 5.5. Bar chart.

POINT AND FIGURE CHARTS

Some chartists use a point and figure chart rather than a bar chart. These chartists maintain that time is not important and only price movements matter, so prices are recorded when they move a certain amount. They record price increases with Xs and price decreases with Os as shown in Figure 5.6. When the price moves past the previous high or low, a bull or bear market signal is given. Point and figure charts also have numerous other formations to help predict price directions such as the ones used in bar charts plus fulcrums, catapults, and others.

MOVING AVERAGES

Moving averages are a very popular way to forecast prices. They are simple to calculate and easy to interpret. Usually at least two averages are used together and oftentimes three. Table 5.2 shows how to calculate 3-, 5-, and 10-day moving averages. Figure 5.7 shows these averages graphically. The forecast rules are also simple: when the short-term average crosses the intermediate average a watch signal is given; when the intermediate average crosses the longer average, a bull or bear market direction is indicated.

Averages are easy to program for a personal computer, so much of the work involved in using moving averages can be performed by a computer or a good calculator.

Point & Figure

```
        X           X
200   X O X O X O
        O X O   O
        O X     O
180   O         O
                O X
160   ·         O X O X
                O   O   O
```

Figure 5.6. Point and figure chart.

Table 5.2. Calculating 3-, 5-, and 10-day moving averages[a]

Daily Price	3-Day Average	5-Day Average	10-Day Average
	--------------------dollars--------------------		
3.00			
3.10			
2.99	3.03		
3.05	3.05		
2.98	3.01	3.02	
3.11	3.05	3.05	
3.15	3.08	3.06	
3.19	3.15	3.10	
3.21	3.18	3.13	
3.22	3.21	3.18	3.10
3.20	3.21	3.19	3.10

[a]The averages are calculated by taking the last 3-, 5-, or 10-day prices, adding them, and dividing by the number of observations for each average. As a new day is added, the oldest is removed. Notice that the longer the time period for the averages, the more stable the average value is relative to the actual prices.

Forward Contracts

A forward contract is simply an agreement between the buyer and seller relating to the delivery and acceptance of a specific commodity at some point in the future. For these contracts to have meaning and be effective, they must be written. Although oral contracts are as binding as written contracts, the actual terms of the contract are subject to individual memories, which at best, are forgetful. When a contract is being entered into, the following items need to be specified:

1. The names of each party and their relationship (especially important in family dealings). Make sure addresses are provided.

2. Specify the commodity or service to be delivered or accepted. If important, specify the quantity, quality, and condition of the product at delivery. If other quality conditions are acceptable, specify premiums and discounts that pertain to each grade or condition.
3. Specify when delivery will take place, where, what form of payment is acceptable, and when title transfers. Specify who will pay the insurance premium if insurance is necessary and who receives the insurance settlement.
4. If price is specified, make sure premiums and discounts for quality variations pertain to the price stated. If a price formula is specified, make sure the formula is stated and the variables used in the formula are stated.
5. If either party has problems with the terms of the contract, specify an arbitration process. This will save both parties from having to resolve differences in court.

This list is not exhaustive, but it provides some of the critical characteristics that good forward contracts must have. All contracts should be reviewed and/or drafted by a competent lawyer. Money spent before the contract is signed is cost-effective.

There are three major types of forward contracts that producers enter into relative to selling their output. They include market specification, production management, and resource providing.

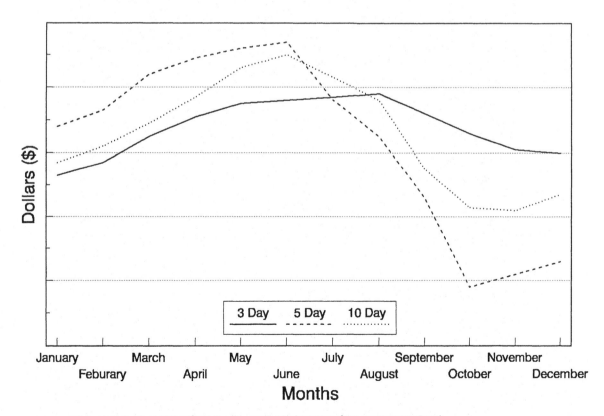

Figure 5.7. Hypothetical 3-, 5-, and 10-day moving average chart.

MARKET SPECIFICATION

A market specification contract is usually entered into after production has begun. It typically provides a known selling point and time for the product. Price or a formula for fixing price is usually specified. Quantity may or may not be specified. The major advantage of this type of forward contract is that the producer has a known selling point for the product. The major disadvantage is that it is entered into after production begins and thus may or may not be available each year. Livestock and grain are the principal types of commodities that are sold using market specification contracts.

PRODUCTION MANAGEMENT

Most processors try to enter production management contracts with producers so they can better control timing of deliveries, quality, and production within the plant. Typically production management contracts are available for many vegetables and some fruits. Production management contracts are entered into *before* production has begun. In fact, the degree of contracting is so strong with certain crops that if a contract is not available, most producers do not risk growing the crop. Title is retained until delivery is completed. Often several management practices or seed varieties are specified by the contracting party to whom the product will be delivered.

RESOURCE PROVIDING

Resource providing contracts are very restrictive. The producer is often providing no inputs other than management and some capital equipment. Almost all inputs and cultural and management practices are specified in the contract. The producer does not retain title. Thus the major advantage to the producer is the absence of many types of risk. However, this also becomes the major disadvantage in that the producer loses almost all control. Much of the poultry and eggs in the United States is produced under resource providing contracts. Table 5.3 shows a comparison of these three types of forward contracts.

Hedging

Hedging is a process involving positions in cash and futures markets simultaneously. The idea of hedging is to offset losses in the cash market with gains in the futures market. This generally works because the cash and futures prices move together. Thus if the cash price falls, then the futures price likewise falls. If a corn producer is properly hedged, i.e., sells a corn futures contract, then when the cash corn price falls the futures position gains value. Hedging can be accomplished using futures contracts and/or options on futures contracts.

Hedging with Futures Contracts

A futures contract is nothing more than a promise to deliver (sell) or accept delivery (buy) of a specific commodity with time of delivery, place of delivery, and quality specified on the contract. The contracts are standardized. Thus all corn futures contracts on the Chicago Board of Trade are for 5,000 bushels of #2 yellow corn deliverable at par to Chicago. The standardization permits the contracts to be traded and retraded. All futures contracts are traded through nonprofit exchanges. If a trader is buying a futures contract, he promises to accept delivery of the commodity

Table 5.3. A comparison between the various types of forward contracts

Contract Terms	Market Specification Contracts	Production Management Contracts	Resource Providing Contracts
Specifies delivery schedule	Yes	Yes	Yes
Signed after production begins	Yes (usually)	No	No
Title of commodity with producer	Yes	Yes	No
Price to producer	Fixed or by formula	Fixed or by formula	Based on volume of output
Specifies quantity	Sometimes	Sometimes	Yes
Specifies quality	No (usually)	Yes	Yes
Specifies cultural practices	No	Some	Yes
Money advanced to producer by contractor	Few	Few	Sometimes
Input supplied by contractor	No	Some	Most
Inputs financed by producer	Yes	Most	Few
Examples	Grain, Cotton Livestock	Vegetables for Processing	Broilers Eggs

Source: Adapted from R.L. Mighell and L.A. Jones, *Vertical Coordination in Agriculture,* Ag Econ Report No. 19, Economic Research Service, U.S. Department of Agriculture, February 1963.

according to the contract specifications. Most producers will be involved in selling futures contracts as hedges. Let's consider both a selling hedge (called a short hedge) and a buying hedge (called a long hedge).

SHORT HEDGE

Short hedges are usually used by producers of a product to protect against decreases in the cash price. Consider a corn producer who has planted corn on May 15 and normally harvests in November. The hedge would look like the example in Table 5.4.

Table 5.4. A corn production hedging example showing both price increases and decreases and the effect on net hedged price

Cash Position	Futures Postition
<u>May 15</u> Crop planted Local cash price $3.50/bu.	Sell December futures @ $3.70

------------------------------Price Decrease------------------------------

<u>Nov. 15</u> Sell crop in local market @$2.50	Buy December futures @$2.70 +$1.00/bu.

Net Hedged Price = $2.50 + $1.00 = $3.50

------------------------------Price Increase------------------------------

<u>Nov. 15</u> Sell crop in local market @ $4.50	Buy December futures @$4.70/ -$1.00/bu.

Net hedged price = $4.50 - $1.00 = $3.50

Notice that whether or not the price went down or up the producer received $3.50 per bushel. The example is a perfect hedge—there was no basis change. Basis is the difference between the cash price for the commodity and the futures price. Obviously, when cash prices decrease the futures will decrease, but usually futures prices will not change by the exact same amount—maybe more, maybe less. Therefore, there is always the possibility of basis changes. These changes will affect the net hedged price. This is illustrated in Table 5.5.

LONG HEDGE

A long hedge is useful for producers when they need to protect against price increases, such as input purchases. Table 5.6 shows an example of a long corn hedge.

TARGET PRICES

Because basis movements and values impact upon the final net hedge price, they can be useful in forecasting net hedged price. An estimate needs to be made of the ending basis value. This is often done by using past basis values. Most land-grant universities have calculated historical basis values that can be used as estimates. A target price is simply a forecast of what the net hedged price will be. It is usually calculated as:

> Futures price at beginning of hedge period
> − Estimate of ending basis
> ─────────
> = Target price

Table 5.5. Basis changes and how they affect net hedged price

Cash Position	Futures Position	Basis
May 15 Crop planted Local cash price $3.50/bu.	Sell December futures @$3.70/bu.	$0.20/bu.

-------------------------------Basis Improvement---

Cash Position	Futures Position	Basis
Nov. 15 Sell crop in local market @ $3.40/lb.	Buy December futures @$3.50 +$.20/bu.	$0.10 Change of $0.10/bu.

Net Hedged Price = $3.40 + $0.20 = $3.60

-------------------------------Basis Deterioration---

Cash Position	Futures Position	Basis
Nov. 15 Sell crop in local market @ $3.40/bu.	Buy December futures @$3.80/bu. -$0.10/bu.	$0.40 Change of $0.10/bu.

Net hedged price = $3.40 - $0.10 = $3.30

Table 5.6. Example of a long hedge with corn futures

Cash Position	Futures Position
Need in two weeks to buy corn for feeding operation. Do not have storage space today. The price today is $3.50/bu.	Buy December futures @ $3.70/bu.

-------------------------------Price Increase-----------------------------------

Two weeks later

Cash Position	Futures Position
Buy corn locally for $3.90/bu.	Sell December futures @$4.10/bu. +$0.40/bu.

Net Hedged Price = $3.90 - $0.40 = $3.50/bu.

-------------------------------Price Decrease-----------------------------------

Two weeks later

Cash Position	Futures Position
Buy corn locally for $3.30/bu.	Sell December futures @$3.50/bu. -$0.20/bu.

Net Hedged price = $3.30 - (- $0.20) = $3.50/bu.

As an example, consider a corn producer who observes that the December corn futures is trading at $3.50/bu. when he plants his crop on May 15. He looks up the historical basis value for November 15 for the December corn futures contract in the basis tables provided by the Cooperative Extension Service. That value is 20 cents. The 20 cents represents the average value of the basis on November 15 for the period evaluated. Using the formula for target prices yields the following:

$3.50 Futures price on May 15
−.20 Estimated ending basis, November 15
———
$3.30 Target price

The target price of $3.30 is the estimated net hedged price that will be received if the producer hedges the crop on May 15, sells in the cash market on November 15, and the basis turns out to be 20 cents.

HEDGING WITH OPTIONS

Options are now available on many futures contracts, with more commodities being added. An option is a right, but not an obligation, to purchase or sell a futures contract. The right to buy is referred to as a call option, and the right to sell is referred to as a put option. If you buy a put option, you purchased the right but not the obligation to sell a futures contract. If you bought a call option, you purchased the right but not the obligation to buy a futures contract. If you sell a call you have obligated yourself to deliver a long futures contract to the buyer, vice versa for a put. Thus only buyers of options, not the sellers, have a right but not an obligation to buy or sell a futures contract. The sellers of options (also called writers or grantors) have the obligation to perform. Sellers must receive some compensation for this obligation. This compensation is called the premium. Buyers of options pay the premium for the right but not the obligation to perform, and sellers receive the premium for taking on the obligation to perform should the option be exercised.

Hedging with options involves only the purchase of options. It does not involve the selling of options. Selling options may prove to be an excellent strategy, but it is not hedging.

Consider the futures hedging example and show how options can be used. Instead of selling a December corn futures contract, we will buy the right, but not the obligation, to sell a futures contract. What we have done with the option is buy the right but not the obligation to hedge. In the case of hedging with futures contracts when prices moved against us (down), the hedge protected us, but when prices moved in our favor (up), we gained in the cash and lost in the futures. With options hedging, this is eliminated, as illustrated in Table 5.7.

Notice that when prices move down, we exercise the option and hedge with futures. However, when prices move up we let the option expire and capture the gain in the cash market. It must be noted also that the option hedge yields a lower net hedged price by the amount of the premium than the futures hedge when prices move against the hedger. But at the same time, when prices move in favor of the hedger, the options hedge outperforms the futures hedge. This gives us the general rule of thumb to determine whether or not to hedge with options or futures: when the probability that prices will move against the hedger is greater than the probability that they will move in favor of the hedger, hedge with futures. When the probability that

Table 5.7. Comparison between a futures short hedge and an options hedge

Cash Position	Futures Position	Options Position
May 15 Crop Planted Local cash price $3.50/bu.	Sell December @ $3.70/bu.	Buy put option Strike price of $3.70/bu., Premium of $0.10

--Price Decrease---

Nov. 15 Sell locally @ $3.00/bu.	Buy December @ $3.20/bu. +$0.50/bu.	Exercise option to Sell December at $3.70/bu. Buy December at $3.20/bu. +$0.50 less $0.10 premium, net + $0.40/bu.

Net hedged price, futures = $3.00 + $0.50 = $3.50/bu.
Net hedged price, options = $3.00 +$ 0.40 = $3.40/bu.

--Price Increase--

Nov. 15 Sell locally @ $4.50/bu.	Buy December @ $4.70/bu. -$1.00/bu.	Let option expire _____ Lose premium of $0.10

Net hedged price, futures = $4.50 - $1.00 = $3.50/bu.
Net hedged price, options = $4.50 - $0.10 = $4.40/bu.

prices will move in favor of the hedger is greater, hedge with options.

Additional considerations concerning options versus futures involve the fact that buying options as a hedge does not require the posting of margins or margin calls. Likewise, the premium is known in advance and is the most the hedger can lose. One major disadvantage of options right now is that the available options are limited in number and are not being offered very far into the future.

OPTION PREMIUM AND STRIKE PRICES

Traders have several choices of which strike price to pick for an option and what level of premium to pay. For most futures contracts, at least seven choices will be available for strike prices each trading day. This gives rise to another set of terms that must be learned. Options are classified according to whether they are in the money, out of the money, or at the money. Whether or not an option is classified by these terms is determined by the strike price relative to the underlying futures price. Thus if the underlying December corn futures is trading at $3.00/bu. and there is a strike price for a put at $3.20, the $3.20 strike price is said to be in the money by 20 cents. Likewise a $3.20 call option would be out of the money. This is most easily seen when viewed as exercising the option. When the $3.20 put is exercised, the trader will be placed short one December corn futures at $3.20. She could then turn around and buy a futures at $3.00 and thus make 20 cents. This 20 cents is called intrinsic value and is the amount the option is in the money. The opposite is true for the $3.20 call.

Obviously, the premium will reflect this profit because no seller would offer a $3.20 put for a premium of anything less than 20 cents.

Having all of these choices gives hedgers a wider set of strategies with which to place hedges. They could, for example, buy cheap out of the money options to protect at a certain level, or decide to pay more in premiums to get a higher net hedge price. Options offer hedgers true price insurance—for the payment of a premium they are protected. However, if the protection is not needed, they benefit from the price movement in their favor.

THE MARKET PLANNING PROCESS: SNAKE BELLY FARMS

Let us go through an example of the market planning process for Joe and Mary Farmer using the input nitrogen fertilizer. (The process is the same whether it is for an input or an output.) A situation analysis is developed first, goals are formulated, and then strategies are devised to achieve the goals.

Joe and Mary first need to determine what they have to work with. An inventory of physical assets shows that the couple does not have any storage for liquid nitrogen nor do they have enough covered storage to handle bagged or bulk fertilizer. They do not have application equipment, because they have always contracted for application. They do have the necessary tractor power to apply the fertilizer if an applicator is available from the supplier. The tractor power would be available from late November to the end of March.

Inventorying management and marketing skills is also part of determining assets. Mary is willing to spend the necessary time to search for price, quantity, and quality information from various sources for comparison purposes. She has knowledge of forward contracts, purchasing, and record keeping. Both Mary and Joe are aware of product differentiation among fertilizer companies and always purchase fertilizer based upon chemical analysis, price, and service.

Joe and Mary then need to look at how they've used fertilizer in the past. An inventory of currently purchased inputs shows that they purchased 38 tons of liquid nitrogen last crop year for application on corn.

An inventory of current purchasing procedures shows that the 38 tons purchased last season were purchased ten days before planting (May 1) under open account with the local supply cooperative (Joe and Mary are members). They paid $100/ton for the fertilizer (due and paid May 31). The cooperative's repayment ratio this year was 20/80 (they paid 20 percent of patronage dividends earned and deferred the remainder). The coop's record for repayment of deferred dividends is poor. To date, Joe and Mary have never been repaid any of the deferred dividends. Thus they do not consider the 80 percent as collectible. They received a check from the coop at year's end for their 20 percent or $38 (the total dividends that they must report as income is $190). The net cost of the fertilizer to Joe and Mary was $99/ton (actually the cost was higher because the $1/ton refund was not received until January 31 of the next year after they paid for the fertilizer May 31 of the previous year, an opportunity cost of $1/ton for 8 months).

The third step in the market planning process is to consider future alternatives. First, the couple needs to inventory potential new enterprises. Joe and Mary plan on adding five acres of carrots that will require an additional three tons of nitrogen.

They then should perform a market channel analysis of current and potential enterprises. Mary has found that the local supply coop is the only local retail outlet for liquid fertilizer that provides applicators. She has found that 40 miles away a large regional supply center has liquid fertilizer, applicators available in the fall only, and discounts for fall purchases of fertilizer of 20 percent.

Finally Joe and Mary must evaluate the pricing tools available for each enterprise. Mary has negotiated with the regional supply center, which is willing to sell her 43 tons of fertilizer via a contract signed by June 15 for delivery in December (Mary added an extra two tons on the advice of the extension agronomist because some of the fertilizer will be lost between December and planting time). The supply center will provide an applicator and price the fertilizer at $80/ton. The local coop will not enter into a forward contract with Mary, but it will provide Mary with the fertilizer in December at $90/ton (the price is *not* guaranteed). Joe and Mary must pick up the fertilizer at the place of business for both firms. Mary estimates the transportation costs to the local coop to be $1.00/ton and a commitment of 10 hours at various times to get the product. Transportation costs for the regional supply center would be $5.00/ton and 20 hours of time commitment. Mary estimates that there is ample time available in December to get the product at either location.

Mary obtained estimates from three outlook reports on fertilizer for fall and next spring. All three reports forecast flat prices for fall and from 3 to 7 percent increases in prices from the current spring for next spring. She plans for a 5 percent increase for next spring.

Goal

Joe and Mary formulated the following goal relative to purchases of nitrogen fertilizer: to lower the overall price per ton paid by 5 percent relative to last year's price. Last year's price was $99. So this year's goal is to pay no more than $94.05/ton.

Both Joe and Mary feel this is a realistic goal consistent with their broader goal of cutting input costs without sacrificing quality of inputs.

Market Strategy

To achieve their goal, Joe and Mary feel the following strategy will get them their stated price objective of $94.05/ton:

Plan 1. Deal for the first time with the regional supply center by signing by June 15 a contract for delivery of the fertilizer in December. Joe and Mary will pay $80/ton for 43 tons plus $5/ton delivery charge for an equivalent 41-ton price of $89.14/ton.

Plan 2. If for some reason the contract does not prove worthy, buy the fertilizer (43 tons) from local coop for $90/ton with a $1/ton delivery charge for an equivalent 41-ton price of $95.44.

Plan 3. If application in December proves impossible, spring purchase will be for 41 tons at an average price estimated to be 5 percent above last spring's. The local coop price is projected to be $103.95/ton (counting a 20 percent return in patronage dividends). The regional supply center's

price was $93/ton last spring, giving a projected price this spring of
$97.65/ton. With delivery charges added the local coop price would be
$104.95 and the regional supply center's price would be $102.65. If
plans 1 and 2 do not work out, then Joe and Mary have decided to
reevaluate the prices in May. If the regional supply center's price is at
least $2/ton less, they will trade with the regional supply center.

Only Plan 1 of Joe and Mary's strategy will achieve their goal. It is highly
probable that the goal can be achieved, because Joe and Mary are entering a forward
contract with the regional supply center and the firm has an excellent reputation. Plan
2 of the strategy gets them close to their goal, but not exactly. It, too, has a high
degree of success, because they have been trading with the local coop for five years.
Plan 3 will not achieve their goal nor even get them close, because it is merely a
default position they are willing to take in the event Plans 1 and 2 don't work.

Notice that in doing a marketing plan, the situation analysis section is critical.
From the situation analysis, goals and most strategies can be developed. An individual
market plan for each input and output is helpful but not essential. Inputs and outputs
could be lumped together, especially in developing the situation analysis. For
example, all fertilizers could be lumped into one market plan and perhaps two outputs
such as corn and grain sorghum. Indeed, all market plans need to be viewed together
in the broad context of making sure a strategy for one does not prevent or hamper the
achieving of a strategy plan for another.

SUMMARY

Market planning is forward analysis that is composed of: 1) situation analysis; 2)
establishment of goals; and 3) development of marketing strategies. Several marketing
tools exist that can be useful in developing a market plan. Those tools consist of, but
are not limited to, market channel analysis, seasonal and cyclic price movements,
Delphi Approach, fundamental and technical price analysis, forward contracts, and
hedging with futures and options.

A good market plan will have a detailed situation analysis. After the situation has
been completed, the manager sets goals and develops strategies to achieve the goals.
Several strategies for each goal are desirable so that the market plan has flexibility.
If several good strategies are developed for each goal, then goal switching and
changing is minimized. This results in a market plan that is more likely to be achieved
rather than abandoned.

RECOMMENDED READINGS

Howard Beerman, "How to Reduce Risk in Farm Pricing," *Agri-Finance*, August 1985.

Robert Branson and Douglas Nowell, *Introduction to Agricultural Marketing*, McGraw-Hill, New York, N.Y., 1983.

Gene Futrell, Ed., *Marketing for Farmers*, Doane-Western, Inc., St. Louis, Mo., 1982.

Perry Kaufman, Ed., *The Concise Handbook of Futures Market*, John Wiley & Sons, New York, N.Y., 1986.

Richard Kohls and Joseph Uhl, *Marketing of Agricultural Products*, 6th ed., Macmillan, New York, N.Y., 1980.

Raymond Leuthold, Joan Junkus, and Jean Cordier, *The Theory and Practice of Future Markets*, Lexington Books, Lexington, Mass., 1989.

Todd Lofton, Ed., *Trading Tactics, A Livestock Futures Anthology*, CME, Chicago, Ill., 1986.

Wayne Purcell, *Agricultural Marketing: Systems, Coordination, Cash and Futures Prices*, Reston, Va. 1979.

Wayne Purcell, *Agricultural Futures and Options*, Macmillan, New York, N.Y., 1991.

V. James Rhodes, *The Agricultural Marketing System*, 3rd ed., John Wiley & Sons, New York, N.Y., 1987.

CHAPTER 6

Production Planning

With realistic marketing opportunities identified and reasonable price projections made, the effort of planning production can begin. Planning production, especially planning for input needs and output expectations, draws heavily on the portion of microeconomic theory usually referred to as the theory of the firm or production economics. The first few steps of the production planning process have actually already been taken. We have already identified the potential commodities by taking stock of land, water, machinery, and climatic resources, by considering the preferences and goals of the farm operator, and by assessing marketing opportunities and limitations. We have also already completed an assessment of input and output marketing and price projections that will enable us to turn to determination of optimal levels of input use and output production.

The production planning activities begin with collecting data and reviewing available research results that shed light on the biological processes of each commodity, especially the relationships between yield and various inputs that are summarized by production functions. Given input and output price projections, production function analysis will provide optimal levels of input use as well as yield expectations. The results of the production function analysis are summarized in enterprise budgets and provide the major information required to develop a whole-farm plan, which, in turn, will provide the list of required purchased inputs necessary for the purchasing schedule.

INPUT-OUTPUT RELATIONSHIPS

The primary way of summarizing the biological-technical relationships between input applications and output response is through the production function. Biological scientists and agricultural engineers exert a great deal of research effort in developing production functions relating yield response to various levels of fertilizer or irrigation water application, to chemical pesticide applications or seeding rates, to crop variety selection, or to ration composition. While much has been learned about yield responses, much more research needs to be conducted to be able to develop and analyze appropriate production functions at the farm level. Furthermore, many of the research results are site specific and must be carefully interpreted and adopted for individual farm or ranch use.[1]

Production Function Analysis

Most research publications report production functions in mathematical form. That mathematical form can be analyzed directly, usually through the use of calculus methods, or it can be rewritten into a discrete series of input applications.

A textbook example of a mathematical production function is shown in equation and graphical form in Figure 6.1a and has been rewritten in discrete form in Figure 6.1c. This production function relates yield response of corn to varying levels of nitrogen fertilizer application and shows a typical shape, in that at a low level of input use, yield is low, but increases at an increasing rate at the beginning. After a brief span, maximum efficiency of input use is realized, and thereafter, yield increases at a decreasing rate. Some production functions are linear, i.e., as input use increases, yield increases at a constant rate. Regardless of whether yield increases are shown to occur at a constant, increasing, or decreasing rate, eventually a yield maximum will be reached. Once the yield maximum is reached, further applications of the input will actually decrease yields as the fertilizer burns the plant, or the water drowns the plant, or the animal exhibits a toxic response to a particular ration element.

Analysis of the production function begins with the calculation of the marginal physical product and average physical product (see Figure 6.1b).[2] The marginal physical product concept shows the average productivity or efficiency of the input at each particular level of input use. Marginal physical product (MPP) is calculated by dividing the change in yield response by the change in input use:

$$\text{MPP} = \frac{\text{Change in yield response}}{\text{Change in input use}} = \frac{\Delta \text{ Total physical product}}{\Delta \text{ Input use}} = \frac{\Delta \text{ TPP}}{\Delta \text{ Input}}$$

Thus, MPP ignores any input level prior to the last increment. Average physical product, on the other hand, relates total physical product or yield response to input use by calculating the average yield per unit of input use, i.e., the efficiency or productivity of the input at that level. Average physical product (APP) is consequently computed by dividing yield response by input use:

$$\text{APP} = \frac{\text{Yield response}}{\text{Input use}} = \frac{\text{TPP}}{\text{Input}}$$

Stages of Production

Interpretation of the relationship between MPP and APP is critical. If MPP is greater than APP, then APP must increase, because if the last unit of input applied is more productive than the average of previous units, then average productivity must increase. Conversely, if MPP is less then APP, then APP must decrease, because if

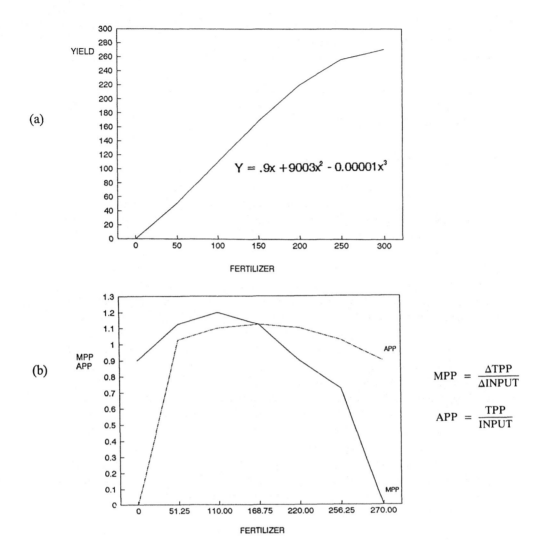

(a)

$$Y = .9x + 9003x^2 - 0.00001x^3$$

(b)

$$MPP = \frac{\Delta TPP}{\Delta INPUT}$$

$$APP = \frac{TPP}{INPUT}$$

(c)

Fertilizer (Pounds)	Yield (Bushels)	MPP	APP
0	0	--	--
50	51.25	1.025	1.025
100	110.00	1.175	1.100
150	168.75	1.175	1.125
200	220.00	1.025	1.025
250	256.25	1.725	1.025
300	270.00	0.275	0.900

Figure 6.1. Example of fertilizer production function for corn.

the last unit of input applied is less productive than the average of previous units, then average productivity must decrease.[3]

At the point where MPP is just equal to APP, an important dividing line exists. The point marks the dividing line between Stage I of the production function and Stage II. Stage I is called irrational, because up to the point where MPP = APP, each additional unit of input makes all previous units more productive or efficient. If the first unit can be justified, i.e., it provokes enough yield response to offset its cost, then the last unit prior to the MPP = APP point must also be cost-effective, because it made the first unit even more efficient. In other words, there is no economic justification for stopping anywhere within Stage I—either no inputs should be applied or at least enough should be applied to reach the point where MPP = APP.[4]

To the left of the MPP = APP point, MPP > APP and thus creates Stage I, but to the right, MPP < APP. In this area, known as Stage II, increasing input use causes APP to fall. It is not possible to make a claim for the rationality of producing at any particular point in Stage II with only the physical information we have so far. But before we bring in the economic information, let's dispense with Stage III. Stage III begins at the point where MPP = 0. Increasing input beyond this level causes MPP to be less than zero, or in other words, TPP or total yield falls. There is certainly no reason to increase inputs and thus incur greater costs if that application causes yield to fall. Consequently, Stage III is clearly irrational just like Stage I.

We are thus left with no input application or somewhere in Stage II as rational. If profitable at all, determination of the point in Stage II that is the most profitable requires the knowledge (or projection) of both input and output price.

Profit-Maximizing Level

The location of the exact point where profits are maximized can be found in one of two ways. Both require that the price of both the input and the output be known or be estimated. One method deals with determining the optimal level of output and the other method deals with determining the optimal level of input. An example, building on the production function example of Figure 6.1, is provided in Figure 6.2.

The process for determining the optimal level of output begins with the calculation of total and average variable costs (fixed costs are irrelevant at this stage of the argument). Both total variable cost (TVC) and average variable cost (AVC) are easy to calculate if the price of the variable input (P_I) has been estimated:

$$TVC = P_I \times INPUT$$
$$AVC = \frac{TVC}{TPP}$$

The next step is to calculate marginal cost (MC):

$$MC = \frac{\Delta TVC}{\Delta TPP}$$

Marginal cost is perhaps the most important of the three cost figures, because it emphasizes the marginality rules of economics—what happens at the margin is the most important event. MC describes the cost attributed to the additional output generated by the last unit of input used.

(a)

INPUT	YIELD	MPP	APP	TVC	AVC	MC	MR
0	0	--	--	0	--	--	--
50	51.25	1.025	1.025	72.50	1.41	1.41	2.00
100	110.00	1.175	1.100	145.00	1.32	1.23	2.00
150	168.75	1.175	1.125	217.50	1.29	1.23	2.00
200	220.00	1.025	1.100	290.00	1.32	1.41	2.00
250	256.00	0.725	1.025	362.50	1.41	2.00	2.00
300	270.00	0.275	0.900	435.00	1.61	5.27	2.00

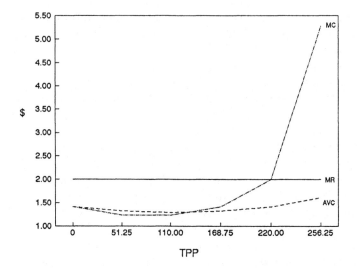

(b)

INPUT	YIELD	MPP	APP	MVP	MIC
0	0	--	--	--	1.45
50	51.25	1.025	1.025	2.05	1.45
100	110.00	1.175	1.100	2.35	1.45
150	168.75	1.175	1.125	2.35	1.45
200	220.00	1.025	1.100	2.05	1.45
250	256.00	0.725	1.025	1.45	1.45
300	270.00	0.275	0.900	0.55	1.45

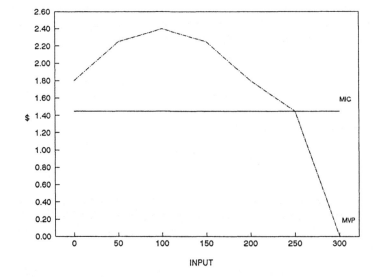

Figure 6.2. Example of fertilizer profit-maximizing levels.

To determine optimality, we need to relate the value of the additional output to its cost. If the last unit of output produces enough additional revenue to pay for the cost of getting it, we will clearly want to apply that last unit. The additional revenue generated by the additional unit of output is called marginal revenue (MR) and is calculated easily if output price (P_o) has been estimated:

$$MR = \frac{\Delta \text{ Total revenue}}{\Delta \text{ Output}} = \frac{\Delta \text{ TR}}{\Delta \text{ TPP}} = \frac{\Delta (\text{TPP} \times P_o)}{\Delta \text{ TPP}}$$

In perfect competition, $MR = P_o$, because an individual producer cannot affect the market output price by increasing or decreasing production. The profit-maximizing level of output occurs at the point where MC = MR. That is, profits are maximized exactly at the level where the last unit of output generates enough revenue to just pay for itself.

The process for determining the optimal level of input is similar to that of the optimal level of output and, if done properly, will produce the same result. The process begins with the calculation of the marginal value product (MVP): $MVP = P_o \times MPP$. MVP is interpreted as the additional value of output generated by an additional unit of input. Similar to previous reasoning, we need to relate this value to the additional cost incurred in applying that input (marginal input cost or MIC). At the point where MVP = MIC, the cost of the last unit of input is equal to the value it generates, and, consequently, profits are maximized.

Multiple Variable Inputs

The reasoning outlined above describes the determination of a profit-maximizing level of output or input as if only one variable input exists. If more than one variable input must be analyzed, as is often the case when both fertilizer and irrigation water are known to affect yields, the analysis can be conducted in a similar, but more complicated, manner. All combinations of the two (or more) inputs should be listed in order of increasing cost. If two combinations result in the same total cost but one has a lower yield, eliminate the one with the lower yield because it is clearly inferior. Then proceed in the same manner as before, considering the inputs as a package of inputs rather than a single input.

Lumpy Inputs

For many inputs, the managerial decision is a simple yes or no rather than the selection of an optimal level from a series of different levels. This situation arises often with respect to pesticides, for example. When applying herbicides or insecticides, it is not often the case that more chemical will increase yield or even more separate chemical applications will increase yield. More common is the case where one application will adequately control the yield-reducing activity of the pest. In this case, the optimal level decision is similar theoretically, but much easier than in the multiple level case. If MC < MR or MVP > MIC, then apply the input. If MC > MR or MVP < MIC, do not apply the input.

Production Function Analysis: Snake Belly Farms

Joe Farmer began his production planning process by contacting his county extension agent. He had already identified the crop and livestock commodities that he was able to produce and market in his area and the commodities he was willing to produce. Once the crop and livestock commodities were identified, he wanted to determine how to maximize the production of each one. To do that, he felt it would be best to review the new research applicable to his area and to consult experts who could add information that might enhance his own knowledge and experience. Joe's county extension agent showed him a new research bulletin reporting the results of a study aimed at finding the yield response of corn, grain sorghum, and wheat. The study was conducted at a nearby university research station. Although this research station does not have exactly the same soil types and growing conditions as Snake Belly Farms, Joe and his county agent felt that conditions were similar enough to apply without significant modification.

The corn production response curve is shown in Figure 6.3. Analyzing this production function using a projected corn price of $2.10 per bushel and pumping costs (variable costs only) of $1.75 per acre-inch, Joe concluded that the profit-maximizing level of production would be about 170 bushels per acre using 48 acre-inches of irrigation water per acre.

Joe computed the optimal irrigation and fertilizer application levels for each of the other crops that he was considering in a similar manner. The optimal fertilizer-irrigation = yield levels for each crop under consideration are shown in Table 6.1.

ENTERPRISE BUDGETING

Production function analysis is the first and possibly most important step in production planning; however, the presentation of production function analysis results

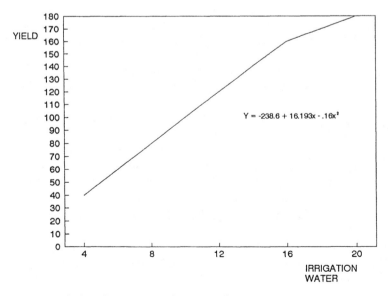

$$Y = -238.6 + 16.193x - .16x^2$$

Figure 6.3. Corn-irrigation production function.

Table 6.1. Optimal fertilizer-irrigation-yield levels for 19X2 crop production on Snake Belly Farms

Crop	Fertilizer (N)	Optimal Level of Irrigation Water	Yield
	pounds	acre-inches	
Alfalfa	20	54	7 tons
Corn	200	48	170 bushels
Cotton	32	28	750 pounds
Soybeans	32	21	45 bushels
Wheat	150	28	70 bushels
Grass Pasture	16	36	14 AUM

is incomplete in the sense that only the variables under consideration are presented. Much more goes into the production process than one or two variable inputs. Other purchased inputs, including pesticides, fuel, labor, baling wire, and machinery operations are necessary. Usually these inputs do not directly affect the yield level if done efficiently and on a timely basis; rather, they usually determine whether any production can be realized. Their necessity is clear, regardless of whether they determine the level of yield, and the fact that these inputs and activities cost money is clear. To prepare an adequate net return figure for production, these costs must also be accounted for.

Enterprise Budget Format

The enterprise budget is a tool that is widely used to show the costs and returns associated with the production of a commodity. Enterprise budgets (a simplified example is shown in Table 6.2) vary in format, but always represent a selected base unit (such as one acre, one head, one hundredweight, or 100 head, for example) and always show expected returns and total production costs. Usually, the returns section shows expected prices (derived from the marketing and price projection activities); expected yields (derived from the production function analysis); other sources of income such as grazing of winter growth or crop aftermath, baling of wheat straw, sales of byproducts like cottonseed, and directly associated government program payments; and the resulting total income. The costs sections usually show purchased input amounts, prices, and costs; input requirements of labor and machinery; other variable costs such as repairs; and fixed costs such as land taxes, depreciation, and overhead. *All* costs associated with the production of the commodity must eventually be shown on the enterprise budget.

Gross Margins

Although enterprise budgets are almost always computed as full-cost estimates, the allocation of fixed costs is a special problem that is somewhat out of order in this discussion. Fixed costs, by definition, do not vary with the level of production and would, in fact, not change if no production at all takes place. These costs (clear examples would be the interest cost on a farm ownership loan or depreciation of

Table 6.2. Condensed enterprise and return estimate: U.S. wheat production, economic costs, and returns per planted acre, 1987-89[1]

Item	1988	1989	1990
		Dollars	
Gross value of production (excluding direct Government payments):			
Wheat	95.89	99.90	94.27
Wheat straw	3.78	3.45	1.52
Total, gross value of production	99.67	103.35	95.79
Economic (full-ownership) costs:			
Variable cash expenses	46.31	53.01	52.64
General farm overhead	6.89	5.01	6.47
Taxes and insurance	8.19	8.72	10.28
Capital replacement	0.67	9.66	9.89
Operating capital[2]	1.78	2.12	1.97
Other nonland capital[3]	4.33	9.67	10.67
Land[4]	31.38	23.27	22.83
Unpaid labor	5.77	8.67	11.24
Subtotal	125.32	120.13	125.99
Residual returns to management and risk[5]	-25.65	-16.78	-30.20
Harvest-period price (dollars per bushel)	3.50	3.81	2.78
Yield (bushels per planted acre)	27.42	26.22	33.91

Source: USDA, *Economic Indicators of the Farm Sector: Costs of Production—Major Field Crops, 1990,* ECIFS 10-4, Agriculture and Rural Economy Division, Economic Research Service, U.S. Department of Agriculture, July 1992.

[1]Includes both operator and landlord costs and returns. Excludes direct government payments and program participation costs.

[2]Variable cash expenses multiplied by part of year used and by six-month U.S. Treasury Bill rate.

[3]Value of machinery and equipment multiplied by long-run rate of return to production assets in the farm sector.

[4]Rental value based on composite share and cash rent.

[5]Gross value of production less total economic costs.

owned machinery) must be allocated to and borne by the commodities produced. However, the allocation process must know how much of each crop is to be produced, something that we don't know yet and won't know until the whole-farm plan has been completed.

Despite the fact that we can't complete the enterprise budget until we determine the whole-farm plan, we need the enterprise budget to be able to do the whole-farm plan. Although it is beginning to sound like we have been caught in a dilemma, we really can proceed without making up numbers. All that is required to develop the whole-farm plan is the gross margin for each commodity, i.e., the return over variable costs. If we can project prices, yields, and variable costs, we can calculate gross margins, proceed to the whole-farm plan, and later return to the allocation of fixed costs.

A negative return for a particular commodity will not keep that commodity out of the whole-farm plan, because that plan is a short-run plan. A commodity will not be produced, even in the short run, if gross margins are negative; but it can be produced in the short run if all costs cannot be covered. However, if commodities with negative returns over all costs are produced, a long-run problem exists. Alternative commodities and alternative marketing strategies with higher returns must be identified, non-productive costs must be eliminated, and more efficient, lower-cost production technologies must be found, or the producer must face eventual exit from farming.

Enterprise Budgets: Snake Belly Farms

After computing optimal yields of each commodity under consideration, Joe Farmer began the process of computing enterprise budgets. He not only listed the fertilizer and irrigation water requirements, but he also listed all of the necessary field operations and estimated their costs. To accomplish this project, he consulted enterprise budgets published for his area by his state agricultural college, modified those budgets based on his farming methods, and prepared his own budgets using a microcomputer budget generator.[5]

The gross margins enterprise budget for corn that Joe developed is shown in Table 6.3. (All of his other budgets will be shown later in completed form.) To get to this stage, however, Joe had to determine all of his purchased input prices (Table 6.4), pumping costs (Table 6.5), equipment needs, and fuel and repair costs (Table 6.6).

THE WHOLE-FARM PLAN

With all of the cost and return information in place, at least through the calculation of gross margins, the next major step is to compute the combination and amounts of commodities to produce to maximize profit. The whole-farm planning process is a long, complicated procedure if done with the goal of maximizing profit *and* if the plan is going to recognize limitations of soil conservation, crop rotations, and other physical constraints. Of course, the process can be estimated by a trial-and-error process of computing total farm gross margins using a series of selected combinations of crops. But in the long run, the most efficient way to proceed is to follow a more formalized procedure to reach the optimum. Two of the more popular and useful techniques of reaching this optimum are programmed budgeting and linear programming.

Table 6.3. Gross margins corn enterprise budget, Snake Belly Farms: corn for grain, flood irrigation, budgeted per acre costs and returns, 19X2; planting dates: April 15-May 1; harvesting dates: October 1-November 1

ITEM	PRICE	YIELD	TOTAL
GROSS RETURNS			
CORN FOR GRAIN	$2.10	170 BU	$357.00
ASCS DEFICIENCY	$1.21	170 BU	$205.70
ASCS DIVERSION	$2.00	24 BU	$48.00
TOTAL			$610.70

PURCHASED INPUTS	PRICE	QUANTITY		PURCHASED INPUTS	TOTAL
SEED	$0.75	32	000	$24.00	$24.00
NITROGEN (N)	$0.25	200	LBS	$50.00	$50.00
PHOSPHATE (P205)	$25.00	1	ACRE	$25.00	$25.00
INSECTICIDE (CUSTOM)		400	DOLLARS	$20.00	$20.00
CROP INSURANCE		48	AC. IN.	—	—
PUMP WATER*					
SUBTOTAL				$119.00	$119.00

PREHARVEST OPERATIONS	POWER UNIT	ACCOMPLISHMENT RATE		PURCHASED INPUTS	FUEL & LUBE	REPAIRS	TOTAL
DISC	130 HP	0.17	HR		$1.17	$1.84	$3.00
PLOW	130 HP	0.48	HR		$3.29	$4.44	$7.74
FLOAT	96 HP	0.16	HR		$0.81	$0.20	$1.02
FERTILIZE	DEALER APPLIED						
LISTER	96 HP	0.18	HR		$0.91	$0.49	$1.40
PRE-IRRIGATE		0.75	HR		$13.97	$0.82	$14.80
CULT & SPRAY	65 HP	0.26	HR		$1.39	$0.77	$2.16
PLANTER	96 HP	0.26	HR		$1.32	$0.61	$1.93
CULTIVATOR (2X)	65 HP	0.42	HR		$2.25	$0.81	$3.05
DITCHER (2X)	96 HP	0.10	HR		$0.51	$0.08	$0.58
IRRIGATE (5X)		2.50	HR		$69.86	$4.12	$73.98
					—	—	—
SUBTOTAL		5.28	HR		$95.48	$14.18	$109.67

HARVEST OPERATIONS				PURCHASED INPUTS			TOTAL
COMBINE (CUSTOM)				$37.40			$37.40
HAUL (CUSTOM)				$18.70			$18.70
				—			—
SUBTOTAL				$56.10			$56.10

OVERHEAD EXPENSES							TOTAL
DOWNTIME		0.51	HR				$0.00
EMPLOYEE BENEFITS							$0.00
INSURANCE				$0.00			$0.00
LAND TAXES							$0.00
SUPERVISION AND MANAGEMENT							$0.00
OTHER EXPENSES				$17.28			$17.28
				—			—
SUBTOTAL		0.51	HR	$17.28	—	—	$17.28

VARIABLE OPERATING EXPENSES		5.79	HR	$192.38	$95.48	$14.18	$302.04
RETURN OVER VARIABLE EXPENSES							$308.66

Table 6.4. Cost information for Snake Belly Farms, 19X2

A. Basic cost information for Snake Belly Farms, 19X2

Item	Unit	Rate
Labor Wage Rate:		
Equipment operators	$/hour	$5.00
General & irrigators	$/hour	$4.00
Purchased Inputs:		
Fertilizer:		
Nitrogen (N)	$/pound	$0.25
Phosphate (P$_2$O$_5$)	$/pound	$0.25
Seed:		
Alfalfa	$/pound	$2.50
Cotton	$/pound	$0.40
Soybeans	$/pound	$0.40
Corn	$/bag	$60.00
Wheat	$/pound	$0.20
Pasture grass	$/pound	$1.00
Plastic Twine:	$/box	$26.00
Natural gas	$/MCF	$4.65
Diesel fuel	$/gallon	$1.00
Gasoline	$/gallon	$1.10
Electricity	cents/KwHr	9.07
LP Gas	$/gallon	$1.05
Employee Liability Insurance	$/wages	$15.00
Employee Benefits	percent/wages	15.00%
Labor Downtime	percent	25.00%
Financial Rates:		12.00%
Operating Capital Interest Rate	percent	12.50%
Land Interest Rate	percent	12.00%
Equipment Interest Rate	percent	4.00%
Real Interest Rate	percent	
Land Taxes	$250.00/acre	$2.24
Personal Property Tax Rate	$/1000 (Assessed Value)	$26.86
Supervision Factors		
Field Crop-Irrigation	$/labor hour	$0.90
Field Crop-Equipment & General	$/labor hour	$0.45
Management Rate	percent	5.00%

B. Overhead cost information for Snake Belly Farms, 19X2

Item	Rate	Cost
Electricity (Domestic & Shop)	$120.00 per month	$1,440
Telephone	$50.00 per month	$600
Accounting & Legal		$1,000
Misc. Supplies & Hand Tools		$1,500
Pickup & Auto		
miles 20000 @	$0.21 per mile	$4,200
Insurance		
general liability (non-employee)		$1,000
fire/theft		$800
Property Taxes		
non-farm land		$150
other than land & machinery		$500
Building repairs and maintenance		$900
Dues, fees, publications		$250
Farmstead Equipment		$100
	Total	$12,440
	Total Per Acre	$17.28

Table 6.5. Pumping costs and data for irrigation wells, Snake Belly Farms, 19X2

INPUT DATA		DEPTH CHARACTERISTICS	
DELIVERY PSI	0	STATIC..................	150
		DRAW DOWN..........	30
		TOTAL HEAD........	180
		CASING................	600
GALLONS PER MINUTE (GPM)...	1000		
WORK HORSEPOWER..............	45		
EFFICIENCY FACTOR:			
ELECTRICITY....................	0.540		
NATURAL GAS....................	0.154		
LP GAS...........................	0.154		
DIESEL...........................	0.160		
FUEL COST PER UNIT:			
ELECTRICITY....................	9.07	¢/KwHr	
NATURAL GAS..................	$4.65	$/MCF	
LP GAS...........................	$1.05	$/GAL	
DIESEL...........................	$1.00	$/GAL	

PUMPING FUEL COSTS

ELECTRIC WELL:	
COST PER HOUR.............$	5.73
COST PER ACRE INCH.....$	2.59
NATURAL GAS WELL:	
COST PER HOUR.............$	3.86
COST PER ACRE INCH.....$	1.75
LP GAS WELL:	
COST PER HOUR.............$	8.62
COST PER ACRE INCH.....$	3.90
DIESEL WELL:	
COST PER HOUR.............$	5.13
COST PER ACRE INCH.....$	2.32

Table 6.6. Equipment summary for Snake Belly Farms, 19X2

ITEM & SIZE	ANNUAL HOURS OF USE	NUMBER	TOTAL VALUE	VARIABLE COSTS				FIXED COSTS		TOTAL PER HOUR
				FUEL AND LUBRICANT	REPAIR	FUEL, LUBE PER HR	REPAIR PER HR	DEPRE-CIATION	TAXES	
TRACTOR 65 HP	248	1	$1,400	$1,326	$202	$5.35	$0.81	$367	$25	$1.58
TRACTOR 96 HP	187	1	$4,300	$951	$135	$5.07	$0.72	$687	$46	$3.91
TRACTOR 130 HP	274	1	$28,000	$1,883	$1,794	$6.86	$6.54	$4,360	$195	$16.60
COTTON PICKER 2-ROW	93	1	$27,500	$562	$1,862	$6.05	$20.02	$3,850	$345	$45.10
SWATHER 14 FT	256	1	$9,200	$958	$965	$3.74	$3.77	$2,900	$130	$11.84
BALER 1-TON	112	1	$27,000		$364		$3.25	$6,500	$291	$60.63
PLANTER 4-ROW	20	1	$3,400		$32		$1.63	$680	$30	$36.43
ROLLING CULT 4-ROW	67	1	$1,500		$74		$1.11	$200	$13	$3.20
DISC OFFSET 14 FT	66	1	$5,750		$283		$4.26	$767	$51	$12.31
DRILL 13 FT	26	1	$2,150		$43		$1.63	$430	$19	$17.16
PLANE 12 FT	32	1	$2,500		$6		$0.20	$333	$22	$11.11
FLOAT 14 FT	12	1	$600		$7		$0.56	$60	$4	$5.34
LISTER 4-ROW	14	1	$1,600		$27		$1.98	$213	$14	$16.86
PLOW (MOLDBOARD) 4-16 IN	96	1	$4,000		$261		$2.72	$533	$36	$5.94
SHREDDER 4-ROW	11	1	$2,500		$4		$0.40	$333	$22	$33.88
COTTON TRAILER	75	3	$3,600		$7		$0.10	$480	$32	$6.83
SPRAYER 12 FT	20	1	$1,375		$20		$1.05	$183	$12	$10.03
FRONT END LOADER (ATTACH)	80	1	$2,550		$147		$1.84	$340	$23	$4.54
SEMI-TRUCK & FLATBED	160	1	$15,000	$840	$1,680	$5.25	$10.50	$3,000	$134	$19.59
RAKES	80	1	$3,400		$201		$2.52	$453	$30	$6.05
V-DITCHER	20	1	$1,250		$1		$0.05	$167	$11	$8.91
NATURAL GAS WELL	13,247	4	$70,000	$51,124	$3,014	$3.86	$0.23	$14,167	$1,522	$1.18
TOTAL			$318,575	$57,645	$11,129			$41,003	$3,010	

Table 6.6. (continued)

ITEM	NEW VALUE	USED VALUE	YEARS LIFE	EQUIP CODE	MAX HOURS	FUEL UNIT/HR	LUBE COEF	ACCUM HOURS	AGE	INTEREST EXPENSE
TRACTOR 65 HP	$5,500	$1,400	15	1	600	4.42	0.10	6,446	26	$330
TRACTOR 96 HP	$10,300	$4,300	15	1	500	4.61	0.10	3,375	18	$618
TRACTOR 130 HP	$43,600	$28,000	10	1	900	6.24	0.10	1,646	6	$2,616
COTTON PICKER 2-ROW	$77,000	$27,500	20	5	500	5.04	0.20	465	5	$4,620
SWATHER 14 FT	$29,000	$9,200	10	2	400	3.12	0.20	1,024	4	$1,740
BALER 1-TON	$65,000	$27,000	10	2	500			336	3	$3,900
PLANTER 4-ROW	$6,800	$3,400	10	8	150			98	5	$408
ROLLING CULT 4-ROW	$3,000	$1,500	15	3	350			467	7	$180
DISC OFFSET 14 FT	$11,500	$5,750	15	3	150			465	7	$690
DRILL 13 FT	$4,300	$2,150	10	8	100			131	5	$258
PLANE 12 FT	$5,000	$2,500	15	4	300			224	7	$300
FLOAT 14 FT	$900	$600	15	3	100			84	7	$54
LISTER 4-ROW	$3,200	$1,600	15	3	200			95	7	$192
PLOW (MOLDBOARD) 4-16 IN	$8,000	$4,000	15	3	300			671	7	$480
SHREDDER 4-ROW	$5,000	$2,500	15	5	200			74	7	$300
COTTON TRAILER	$2,400	$1,200	15	4	30			250	10	$144
SPRAYER 12 FT	$2,750	$1,375	15	8	500			137	7	$165
FRONT END LOADER (ATTACH)	$5,100	$2,550	15	3	300			560	7	$306
SEMI-TRUCK & FLATBED	$30,000	$15,000	10	5	300	5.00	0.05	800	5	$1,800
RAKES	$6,800	$3,400	15	6	300			560	7	$408
V-DITCHER	$2,500	$1,250	15	4	300			140	7	$150
NATURAL GAS WELL	$85,000	$42,500	24	$0.23	4000		0.05	3,312	7	$5,100

Programmed Budgeting

According to Donald C. Huffman,[6]

Programmed budgeting is a technique for systematically choosing which alternatives and how many units of each alternative to include in a farm plan. The programming procedure is not a substitute for the budgeting process. Rather, it is a technique for adding rigor and precision to the budgeting processes. The mechanics of the programming process eliminates the trial, and consequently much of the error, of budgeting. The programming process begins with a set of enterprise (or activity) budgets and ends with an optimal combination of enterprises for the specified resource situation. The programming technique is integrated into a complete planning process for the development of a complete farm plan.

The objective of farm planning is to develop an optimum plan that maximizes net returns from the entire "bundle" of resources considering the production alternatives, which are consistent with the manager's goals and abilities. To achieve this, the objective (or criterion) at each stage of the "programming" process is to select the enterprise which yields the greatest "net return" per unit of unused resource.

The initial selection of enterprises is done on the basis of net returns per unit for the most restrictive resource. Adjustments in the initial selection of enterprises are made using the product substitution principle when other resources become more restrictive.

In most cases, using the land-use approach (maximizing returns per acre of land) will minimize the number of steps required in the programming process. Thus, returns per acre of land become the criterion for the initial selection of enterprises. It should be kept in mind, however, that labor or capital may be the most restrictive resource for a particular farm business. If it can be determined that one of these or some other factor is more restrictive than land, then the criterion should be chosen accordingly. The relative abundance of land, labor, and capital in relation to the requirements for these resources by the production alternatives being considered determine the most restrictive resource. The final solution will be the same regardless of which resource is used as the criterion for the initial selection of enterprises, but the number of steps required to obtain the final solution will be minimized if the most restrictive resource is used as the starting basis. *In most cases it is easier to develop enterprise budgets for alternatives adapted to the land-use criterion.* The land-use criterion will be used for outlining the steps in the planning example.

Enterprises are added to the farm plan until no enterprise can be added from the remaining unused resources and no improvement in net returns can be made by substitution. After the enterprises and number of units of each enterprise have been determined, a financial summary is prepared for the complete farm plan including fixed costs as well as variable costs.

Accuracy of the final farm plan is dependent upon:

    (1) selection of applicable alternatives,
    (2) specification of alternatives in an appropriate manner,
    (3) identification of limiting resources,
    (4) appropriate specification of resource restrictions, and
    (5) validity of enterprise budgets developed in terms of
        (a) technical input-output relationships assumed,
        (b) prices assumed,
        (c) technology assumed, and
        (d) inclusion of appropriate resource requirements in the enterprise budgets.

The process of computing the optimal combination of commodities is very structured and concentrates on the value of gross margins per unit of each resource. Again according to Huffman,[7]

The optimum combination of enterprises is determined by means of two work tables developed simultaneously. The primary work table (Figure 6.6) includes a column to identify the enterprises included in the plan, a column to show how many units of each enterprise are included, one column *each* of the resource restrictions, and a column to show the net returns.

Again, to conserve time, the same column headings may be utilized for both tables in Figures 6.6 and 6.7. However, precaution must be taken to leave adequate space for the table in Figure 6.6, because it is not known at this time how many rows will be required. It isn't mandatory that the table in Figure 6.7 be developed as a formal table. However, this permits the planner to keep calculations organized so he may readily check back in case of error and may avoid duplication of calculations.

The following steps are used in the development of the tables in Figures 6.6 and 6.7 to determine the optimum combination of enterprises to be included in the farm plan.

Step 1.   Enter in Row 1 of the table in Figure 6.6 the maximum amount of resources available as shown in the table in Figure 6.4.

Step 2.   In the table in Figure 6.5 find the enterprise that has the *largest* returns per acre of land. Using the resource requirements shown in Figure 6.4 for this enterprise and the quantity of unused resources shown in Figure 6.6, determine the maximum number of units of this enterprise that can be produced. This is done by dividing the quantity of each resource required to produce one unit of the enterprise (Figure 6.4). Enter the quotients in Row 1 of the table in Figure 6.7 and circle the *smallest* number.

Step 3.   Multiply the resource requirements shown in Figure 6.4 by the *smallest* number in the row of Figure 6.7 just calculated and enter the products in Row 2 of the table in Figure 6.6. Multiply the net returns for this enterprise shown in Figure 6.4 by this same number and enter in the net returns column.

Step 4. In the table in Figure 6.6, subtract Row 2 from Row 1 and enter the difference in Row 3. (The net returns column will be added to give the total returns from all enterprises included in the plan.)

Step 5. In Figure 6.5 find the enterprise having the second largest returns per acre of land and repeat Steps 2, 3, and 4. Continue this process until no additional enterprises can be added from the remaining resources shown in the last computed row of the table in Figure 6.6.

Step 6. In Figure 6.5, check the returns per unit of resource used for the resource, which was used up by the last enterprise added. If the returns per unit of this resource for the last enterprise added is *less* than the returns per unit of this resource for all previously included enterprises, the optimum plan has been achieved. If this number is larger than the returns for some previously included enterprise, proceed to Step 7.

Step 7. Divide the last exhausted resource requirement of the enterprise having lower returns per unit of resource by the resource requirement of the last included enterprise. Multiply this number by the net returns per unit for the last enterprise added. If the resultant substitution value is *less* than the net returns per unit for the previously included enterprise, the optimum plan has been achieved. If the resultant value is larger, the previously included enterprise should be reduced by sufficient number of units to allow increasing the size of the last added enterprise until some other resource becomes limiting. This is accomplished in two steps by "adding in" the resource requirements of the enterprise being reduced in size and "subtracting out" the requirements for the additional units of the enterprise being increased in size.

Step 8. Repeat the check described in Step 6 after making the substitution. If land was truly the most restrictive resource, no further checks are necessary. However, to insure this, it is wise to check the returns per unit of each resource exhausted for all enterprises not included in the plan with those included in the plan. Any *excluded* enterprise having higher returns per unit of *exhausted* resource than an included enterprise that can be added from resources released by the included enterprise plus remaining unused resources should be added if it meets conditions described in Step 7. It is quite important to note that the included enterprise must release sufficient quantities of *all exhausted* resources, and the ratio of resource requirements applicable in computing the substitution value is the smallest of *all exhausted* resources rather than the particular resource having the higher return. (pp. 20-21)[8]

Linear Programming

Linear programming is a bit more sophisticated mathematically than programmed budgeting and much easier to computerize because of a wide variety of available computer programs that are capable of solving linear programs in a specified algorithmic method.[9] The drawback to linear programming (LP) is that the programmer/modeler/budgeter must be able to reason all activities in strict algebraic logic and interpret the computer's results. Although this mathematical rigor is frightening to most businessmen, if the whole-farm planning process is going to be done every year, linear programming is by far the most efficient solution method since the model only needs to be developed once and then will accept modifications in price and yield very easily for use in subsequent years.[10]

Resources	Amount Available	Cotton Mixed Soil 1 Acre	Corn Mixed Soil 1 Acre	Corn Clay Soil 1 Acre	Oats Mixed Soil 1 Acre	Oats Clay Soil 1 Acre	Soybeans Mixed Soil 1 Acre	Soybeans Clay Soil 1 Acre	Beef Cattle 1 Cow	Market Hogs 1 Sow
Total Land	200 acres	1.0	1.0	1.0	1.0	1.0	1.0	1.0	2.5	0.5
Mixed Soil Cropland	100 acres	1.0	1.0		1.0	1.0				
Clay Soil Cropland	50 acres			1.0		1.0		1.0		
Cotton Allotment	0 acres	1.0								
April-May Labor	800 hours	15.3	1.6	2.0	0.3	0.3	1.6	1.6	4.0	12.0
Sept-Oct Labor	800 hours	3.0	1.5	1.4	1.5	1.5	0.3	0.3	2.0	12.0
Investment Capital	$10,000								200.0	100.0
Hog Facilities	20 sows									1.0
Net Returns		$81.19	$53.88	$34.59	$15.66	$ 9.16	$28.10	$20.10	$29.16	$136.40

Figure 6.4. Programmed budgeting work table 1.

Resources	Cotton Mixed Soil 1 Acre	Corn Mixed Soil 1 Acre	Corn Clay Soil 1 Acre	Oats Mixed Soil 1 Acre	Oats Clay Soil 1 Acre	Soybeans Mixed Soil 1 Acre	Soybeans Clay Soil 1 Acre	Beef Cattle 1 Cow	Market Hogs 1 Sow
Total Land	$81.19	$53.88	$34.59	$15.66	$ 9.16	$28.10	$20.10	$11.66	$272.80
Mixed Soil Cropland	81.19	53.88	15.66			28.10			
Clay Soil Cropland			34.59		9.16				
Cotton Allotment	81.19								
April-May Labor	5.31	33.67	17.29	52.20	30.53	17.56	12.56	7.29	11.37
September-October Labor	27.06	35.92	24.71	10.44	6.11	93.67	67.00	14.58	11.37
Investment Capital								0.15	1.36
Hog Facilities									$136.40

Figure 6.5. Programmed budgeting work table 2.

Enterprise	No. of units	Total Land	Mixed Soil Cropland	Clay Soil Cropland	Cotton Allotment	April-May Labor	Sept.-Oct. Labor	Investment Capital	Hog Facilities	Net Returns
Resources Available		200	100	50	20	800	800	10000	20	
Market Hogs	20 Sows	10					240	2000	20	2728.00
Resources Available		190	100	50	20	560	560	8000	0	2728.00
Cotton, Mixed Soil	20 acres	20	20		20	360	306			1623.80
Resources Available		170	80	50	0	254	500	8000	0	4351.80
Corn, Mixed Soil	80 acres	80	80			128	120			4310.40
Resources Available		90	0	50	0	126	380	8000	0	8662.20
Corn, Clay Soil	50 acres	50		50		100	70			1729.50
Resources Available		40	0	0	0	26	310	8000	0	10391.70
Beef Cattle	6 cows	15				24	12	1200		174.96
Resources Available		25	0	0	0	22	298	6800	0	10566.66
Cotton, Mixed Soil (Reduce)	3 acres	3	3		3	46	9			243.57
Resources Available		28	3	0	3	48	307	6800	0	10323.09
Beef Cattle	10 cows	25				40	20	2000		291.60
Resources Available		3	3	0	3	8	298	4800	0	10614.69
Corn, Mixed Soil	3 acres	3				5	5			161.64
Resources Available		0	3	0	3	3	282	4800	0	10776.33

Figure 6.6. Programmed budgeting work table 3.

	Total Land	Mixed Soil Cropland	Clay Soil Cropland	Cotton Allotment	April-May Labor	Sept.-Oct. Labor	Investment Capital	Hog Facilities
Market Hogs	400				66	66	100	20
Cotton, Mixed Soil	100	100		20	36	187		
Corn, Mixed Soil	170	80			158	333		
Corn, Clay Soil	90		50		79	271		
Beef Cattle	16				6	155	40	

Figure 6.7. Programmed budgeting work table 4.

Optimal Whole-Farm Plan: Snake Belly Farms

Joe Farmer chose to consult a local agricultural economics graduate student to develop a linear programming model for Snake Belly Farms in order to determine the optimal combination of resources and enterprises. Joe provided the student with all of his marketing plans and gross margins budgets and a list of all of the constraints to production that Joe felt were relevant. Because the physical constraints of Snake Belly Farms were discussed in Chapter 2, we will present them again only in outline form in Table 6.7. Joe, however, did feel that it was necessary to emphasize that Class II land could only be planted to drilled crops because of soil erodibility and productive capacity and that Class III land could only be used for permanent grass pasture. Furthermore, Joe felt that he must maintain a rotation on Class I land that would limit his options on land that was taken out of a long-term stand of alfalfa. At least for one year after the alfalfa was broken out, cotton, corn, or soybeans must be planted on Class I land and either grass pasture or wheat must be planted on Class II land. Finally, Joe specified resource availability levels on labor, available cash, and credit in addition to his land resources, and required that all short-term borrowing be repaid within the production year and that sufficient cash be generated to pay family living expenses and to service long-term debt.

The linear programming modeling effort provided a crop/livestock enterprise that maximized profits and made the most efficient use of Joe's resources. Those results are shown in Table 6.8.

WHOLE-FARM PLAN SUMMARY

The whole-farm plan presents the optimal activity in physical terms, i.e., number of acres or number of head. For final presentation, the plan must be summarized into financial and economic terms as well. First, the summary should be presented as a whole-farm budget, which is nothing more than a projected income statement

Table 6.7. Resource availability limitations, Snake Belly Farms

Resource	Available	Comments
Class I land	400 acres	Suitable for row crops
Class II land	200 acres	Drilled crops only
Class III land	100 acres	Grass pasture only
Buildings		
grain bins	1,500 bushels	Dilapidated, should be removed
grain bins	1,000 bushels	Excellent condition
cattle barns		Very good condition; sufficient for 700 head
Fencing		Good condition; adequate for 700 head
Labor	3,000 hours	Full-time hired worker
	4,000 hours	Operator and family labor
Management		Sufficient for all but fruit and vegetable crops

Table 6.8. Optimal enterprise combination, Snake Belly Farms

Crop	Acres		
	Class I	Class II	Class III
Alfalfa I	300		
Alfalfa II		20	
Corn	0		
Cotton	100		
Pasture II		100	
Pasture III			100
Soybeans	0		
Wheat		80	
Total	400	200	100

Livestock	Head
Stocker steers	540
Beef cows	0

assuming no inventory adjustments. An example whole-farm budget for Snake Belly Farms is presented in Table 6.9. Second, the optimal production levels should be carried back to the enterprise budgets to allow the allocation of fixed and overhead costs to the individual enterprises. To be most accurate, fixed and overhead costs should be allocated to each commodity based on how much of each resource was used by that commodity. For example, the wheat enterprise should not be charged any depreciation on haying equipment or the cotton picker. However, if a tractor was used for twice as many hours on cotton as it was on wheat, cotton should be charged twice as much tractor depreciation as wheat. The entire set of enterprise budgets that Joe developed are presented in Table 6.10 (at the end of the chapter).

SUMMARY

Production planning is an intense, difficult process because of all of the variables that must be considered. Much of the intensity comes from the desire to maximize the potential of the resources committed to the farm or ranch operation. The first step in the production planning process, assuming that input and output market planning has already taken place, is production function analysis. Production functions for each commodity must be located, interpreted, translated to fit local conditions, and analyzed mathematically to find the optimal level of production. The analysis process seeks the input-output level at which MC = MR or MIC = MVP, that is the level where the last unit of output produced just pays for itself. This level will always be found in Stage II of the production function, or at zero if the input is not limited by some outside constraint.

Gross margins enterprise budgets should be developed in order to begin the optimal whole-farm planning process. The optimal whole-farm plan can be developed using one of a number of mathematical techniques including programmed budgeting and linear programming.

The optimal whole-farm plan should be summarized into a whole-farm budget, and full-cost enterprise budgets should be developed, allocating fixed and overhead costs on a use basis.

Table 6.9. Whole-farm summary, Snake Belly Farms, 19X2

GROSS RETURNS				
ALFALFA HAY I	300 ACRES	$182,700		
ALFALFA HAY II	20 ACRES	$1,697		
PASTURE II	100 ACRES	$16,800		
PASTURE III	100 ACRES	$15,240		
WHEAT	58 ACRES			
CROP		$9,338		
GRAZING		$9,698		
ASCS DEFICIENCY		$8,120		
UPLAND COTTON (PICKER)	75 ACRES	$33,750		
COTTON LINT		$2,700		
COTTON SEED		$15,272		
ASCS DEFICIENCY				
STOCKER CATTLE		$247,752		
GROSS RETURN				$543,067
CASH OPERATING EXPENSES				
SEED		$6,160		
FERTILIZER		$18,811		
CHEMICALS		$12,849		
CROP INSURANCE		$1,500		
OTHER PURCHASED INPUTS		$2,431		
CANAL WATER		$0		
FUEL, OIL & LUBRICANTS-EQUIPMENT		$6,521		
FUEL-IRRIGATION		$51,124		
REPAIRS		$11,129		
CUSTOM CHARGES		$9,387		
LAND TAXES		$1,462		
OTHER EXPENSES		$11,488		
FEED		$37,306		
STOCKER CATTLE		$170,100		
LIVESTOCK EXPENSE		$7,290		
HIRED LABOR		$12,000		
TOTAL CASH EXPENSES			$359,558	
RETURN OVER CASH EXPENSES				$183,509
FIXED EXPENSES		$64,002		
TOTAL EXPENSES			$423,560	
NET FARM INCOME				$119,506
LABOR AND MANAGEMENT COSTS		$23,322		
NET OPERATING PROFIT				$96,184
CAPITAL COSTS				
INTEREST ON OPERATING CAPITAL		$3,966		
INTEREST ON EQUIPMENT INVESTMENT		$17,483		
TOTAL CAPITAL COSTS			$21,499	
RETURN TO LAND AND RISK				$74,735

LAND VALUE	RETURN TO RISK*	RETURN ON INVESTMENT**
$500/ACRE	$61,935	15.06%
$750/ACRE	$55,535	12.04%
$1,000/ACRE	$49,135	10.03%
$1,250/ACRE	$42,735	8.60%

*Return to land and risk − (interest rate × land value × acreage)
**Net operating profit ÷ (machinery and equipment value + land value)

NOTES

1. Finding research results appropriate for a specific area can be a difficult task. The county extension office or a nearby agricultural experiment station are the best places to begin the search process. The professionals who work for the extension service and experiment stations not only have access to most, if not all, of the valuable research publications that relate to their area, but they are also trained to diagnose, recommend, and interpret research results. They also have access to the research scientists who develop the background results.

2. Excellent, thorough, and more detailed discussions of production function analysis can be found in almost every farm management textbook. Especially good discussions may be found in R.D. Kay and W.E. Edwards or in E.N. Castle, M.H. Becker, and A.G. Nelson. The reader who is unfamiliar with production function analysis is encouraged to consult one of these texts for a more detailed development than this quick review provides. See recommended reading list at end of Chapter 2 (Kay and Edwards) and Chapter 6 (Castle, Becker, and Nelson).

3. A popular illustration of this marginal-average rule leaves agriculture and economics for a more easily digested form of entertainment: baseball. Assume a baseball player has a batting average of .300 before today's game. If he goes 3 for 5 today, what will happen to his batting average? It will increase, because MPP = .600 > APP = .300. Conversely, if he goes 1 for 5 today, his batting average will fall, because MPP = .200 < APP = .300.

4. This rule can be broken only if there is some outside limitation on the availability of the input that cannot be justified solely on the basis of economic efficiency. Usually these limitations are institutional rather than economic constraints. For example, in most areas of New Mexico, a farmer is not allowed, by law, to pump more than 3.5 acre-feet of irrigation water per acre of water rights.

5. Enterprise budgets can be computed by hand, but the number of calculations required can become very tedious. A standardized budget format and a great number of calculations suggest a perfect application of a computer. Many state extension services have a computerized budget generator for use at home or in the county agent's office.

6. D.C. Huffman, *Programmed Budgeting—A Tool for Complete Farm Budgeting,* A.E.A. Information Series Number 2, Louisiana State University, Baton Rouge, June 1965, pp. 3-5. Reprinted with permission.

7. Ibid., pp. 4-5. Refer to Figures 6.4 through 6.7.

8. Ibid., pp. 20-21.

9. An excellent example of an affordable microcomputer program capable of solving linear programming problems is G.H. Pfeiffer, MPS-PC Linear Programming System, Research Corporation, 1984.

10. Because of its structured algorithmic nature, we will not dwell on linear programming in this book. An interested reader should consult J.T. Scott, Jr., *The Basics of Linear Programming and Its Use in Farm Management,* AET-3-70, Department of Agricultural Economics, University of Illinois, Urbana, October 1970.

RECOMMENDED READINGS

Emery N. Castle, Manning H. Becker, and A. Gene Nelson, *Farm Business Management: The Decision-Making Process,* 3rd ed., Macmillan, New York, N.Y., 1987.

George H. Pfeiffer, *MPS-PC Linear Programming System,* Research Corporation, Tucson, Ariz., 1984.

Robert B. Schwart, J. Milt Holcomb, and Allan G. Mueller, *Mechanics of Farm Financial Planning,* Circular 1042, Cooperative Extension Service, University of Illinois, Urbana, October 1971.

Ronald F. Eickhorst, Duane E. Erikson, and John T. Scott, Jr., *Combining Linear Programming, Cash Flow Analysis, and Counseling to Improve Individual Farm Planning,* AERR 115, Department of Agricultural Economics, University of Illinois, Urbana, December 1971.

John T. Scott, Jr., *The Basics of Linear Programming and Their Use in Farm Management,* AET-3-70, Department of Agricultural Economics, University of Illinois, Urbana, October 1970.

Timothy G. Baker and John R. Brake, *Cash Flow Analysis of the Farm Business,* Extension Bulletin E-911, Cooperative Extension Service, Michigan State University, East Lansing, Mich., October 1975.

Ralph E. Hepp, *Annual Farm Planning,* Agricultural Economics Staff Paper 80-78, Department of Agricultural Economics, Michigan State University, East Lansing, Mich., 1980.

Table 6.10. Enterprise budgets, Snake Belly Farms

A. Alfalfa establishment, Class I land, flood irrigation, budgeted per acre costs and returns for Snake Belly Farms, 19X2; planting dates: September 1-September 31

PURCHASED INPUTS	PRICE	QUANTITY		PURCHASED INPUTS	TOTAL
SEED	$2.50	30	LBS	$75.00	$75.00
INSECTICIDE (CUSTOM)	$10.00	1	ACRE	$10.00	$10.00
PUMP WATER*		16	AC. IN.	—	—
SUBTOTAL				$85.00	$85.00

PREHARVEST OPERATIONS	POWER UNIT		ACCOMPLISHMENT RATE		PURCHASED INPUTS	LABOR	FUEL & LUBE	REPAIRS	FIXED COSTS	TOTAL
DISC	130	HP	0.17	HR		$0.85	$1.17	$1.84	$4.91	$8.77
PLOW	130	HP	0.48	HR		$2.40	$3.29	$4.44	$10.82	$20.96
DISC (2X)	130	HP	0.34	HR		$1.70	$2.33	$3.67	$9.83	$17.53
PLANE (2X)	96	HP	0.48	HR		$2.40	$2.43	$0.44	$7.21	$12.49
DRILL	65	HP	0.21	HR		$1.05	$1.12	$0.51	$3.93	$6.62
DITCHER (2X)	96	HP	0.10	HR		$0.50	$0.51	$0.08	$1.28	$2.37
IRRIGATE (2X)	96	HP	1.00	HR		$4.00	$27.95	$1.65	$8.58	$42.17
SUBTOTAL			2.78	HR		$12.90	$38.80	$12.63	$46.56	$110.90

OVERHEAD EXPENSES	ACCOMPLISHMENT RATE		PURCHASED INPUTS	LABOR	FUEL & LUBE	REPAIRS	FIXED COSTS	TOTAL
DOWNTIME	0.45	HR		$2.23				$2.23
EMPLOYEE BENEFITS				$1.94				$1.94
INSURANCE			$0.19					$0.19
LAND TAXES				$11.45				$11.45
SUPERVISION AND MANAGEMENT								
SUBTOTAL	0.45	HR	$0.15	$15.61	—	—	—	$15.80

TOTAL OPERATING EXPENSES	3.22	HR	$85.19	$28.51	$38.80	$12.63	$46.56	$211.70

*Pump water costs are shown under irrigation in the preharvest operations section.

Table 6.10. (continued).

B. Alfalfa hay, Class I land, flood irrigation, budgeted per acre costs and returns for Snake Belly Farms, 19X2; harvesting dates: May 20-October 15

ITEM	PRICE			YIELD					TOTAL
GROSS RETURNS									
ALFALFA HAY	$87.00			7.00 TONS (DELIVERED)					$609.00
TOTAL									$609.00

PURCHASED INPUTS	PRICE	QUANTITY		PURCHASED INPUTS				FIXED COSTS	TOTAL
FERTILIZER (10-26-10)	$0.20	200	LBS	$40.00					$40.00
INSECTICIDE (CUSTOM)	$20.00	1	ACRE	$20.00					$20.00
HERBICIDE (CUSTOM)	$6.00	1	ACRE	$6.00					$6.00
PLASTIC TWINE	$26.00	170	FT/TON	$7.74					$7.74
ESTABLISHMENT: PRINCIPAL		6	YEARS					$35.28	$35.28
INTEREST								$12.64	$12.64
PUMP WATER*		54	AC. IN.	—				—	—
SUBTOTAL				$73.74				$47.93	$121.66

PREHARVEST OPERATIONS	POWER UNIT	ACCOMPLISHMENT RATE		PURCHASED INPUTS	LABOR	FUEL & LUBE	REPAIRS	FIXED COSTS	TOTAL
IRRIGATE (6X)		3.00	HR		$12.00	$94.32	$5.56	$28.94	$140.82
SUBTOTAL		3.00	HR		$12.00	$94.32	$5.56	$28.94	$140.82

HARVEST OPERATIONS									
SWATHER (5X)	14 FT	0.80	HR		$4.00	$3.00	$3.02	$9.47	$19.48
RAKE (5X)	65 HP	0.25	HR		$1.25	$1.34	$0.83	$1.91	$5.33
BALER	130 HP	0.35	HR		$1.75	$2.40	$3.43	$27.03	$34.61
FRONT END LOADER (5X)	96 HP	0.25	HR		$1.25	$1.27	$0.64	$2.11	$5.27
HAUL	SEMI	0.50	HR		$2.50	$2.63	$5.25	$9.79	$20.17
SUBTOTAL		2.15	HR		$10.75	$10.63	$13.16	$50.31	$84.86

OVERHEAD EXPENSES									
DOWNTIME		0.54	HR		$2.69				$2.69
EMPLOYEE BENEFITS					$3.41				$3.41
INSURANCE				$0.34					$0.34
LAND TAXES								$2.24	$2.24
SUPERVISION AND MANAGEMENT					$34.12				$34.12
OTHER EXPENSES				$17.28					$17.28
SUBTOTAL		0.54	HR	$17.62	$40.22			$2.24	$60.07
TOTAL OPERATING EXPENSES		5.69	HR	$91.35	$62.97	$104.94	$18.72	$129.42	$407.41

NET OPERATING PROFIT									$201.59
INTEREST ON OPERATING CAPITAL	(63.82 @ 12.00%)								$7.66
INTEREST ON EQUIPMENT INVESTMENT									39.38
RETURN TO LAND AND RISK									$154.55

BUDGET SUMMARY

GROSS RETURN		$609.00	
VARIABLE OPERATING EXPENSES	$215.02		
RETURN OVER VARIABLE EXPENSES		$393.98	(GROSS MARGIN)
FIXED EXPENSES	$129.42		
NET FARM INCOME		$264.56	(RETURN TO CAPITAL, LABOR, LAND & RISK)
LABOR AND MANAGEMENT COSTS	$62.97		
NET OPERATING PROFIT		$201.59	(RETURN TO CAPITAL, LAND & RISK)
CAPITAL COSTS	$47.04		
RETURN TO LAND AND RISK		$154.55	

*Pump water costs are shown under irrigation in the preharvest operations section.

Table 6.10. (continued).

C. Alfalfa establishment, Class II land, flood irrigation, budgeted per acre costs and returns for Snake Belly Farms, 19X2; planting dates: September 1-September 31

PURCHASED INPUTS	PRICE	QUANTITY		PURCHASED INPUTS	TOTAL
SEED	$2.50	30	LBS	$75.00	$75.00
INSECTICIDE (CUSTOM)	$10.00	1	ACRE	$10.00	$10.00
PUMP WATER*		16	AC. IN.	—	—
SUBTOTAL				$85.00	$85.00

PREHARVEST OPERATIONS	POWER UNIT		ACCOMPLISHMENT RATE		PURCHASED INPUTS	LABOR	FUEL & LUBE	REPAIRS	FIXED COSTS	TOTAL
DISC	130	HP	0.17	HR		$0.85	$1.17	$1.84	$4.91	$8.77
PLOW	130	HP	0.48	HR		$2.40	$3.29	$4.44	$10.82	$20.96
DISC (2X)	130	HP	0.34	HR		$1.70	$2.33	$3.67	$9.83	$17.53
PLANE (2X)	96	HP	0.48	HR		$2.40	$2.43	$0.44	$7.21	$12.49
DRILL	65	HP	0.21	HR		$1.05	$1.12	$0.51	$3.93	$6.62
DITCHER (2X)	96	HP	0.10	HR		$0.50	$0.51	$0.08	$1.28	$2.37
IRRIGATE (2X)	96	HP	1.00	HR		$4.00	$27.95	$1.65	$8.58	$42.17
SUBTOTAL			2.78	HR		$12.90	$38.80	$12.63	$46.56	$110.90

OVERHEAD EXPENSES					PURCHASED INPUTS	LABOR	FUEL & LUBE	REPAIRS	FIXED COSTS	TOTAL
DOWNTIME			0.45	HR		$2.23				$2.23
EMPLOYEE BENEFITS						$1.94				$1.94
INSURANCE					$0.19					$0.19
LAND TAXES						$11.45				$11.45
SUPERVISION AND MANAGEMENT										
SUBTOTAL			0.45	HR	$0.19	$15.61	—	—	—	$15.80
TOTAL OPERATING EXPENSES			3.22	HR	$85.19	$28.51	$38.80	$12.63	$46.56	$211.70

*Pump water costs are shown under irrigation in the preharvest operations section.

Table 6.10. (*continued*).

D. Alfalfa hay, Class II land, flood irrigation, budgeted per acre costs and returns for Snake Belly Farms, 19X2; harvesting dates: May 20-October 15

ITEM	PRICE	YIELD		TOTAL
GROSS RETURNS				
ALFALFA HAY	$87.00	5.00 TONS (DELIVERED)		$435.00
TOTAL				$435.00

PURCHASED INPUTS	PRICE	QUANTITY		PURCHASED INPUTS	FIXED COSTS	TOTAL
FERTILIZER (10-26-10)	$0.20	160	LBS	$32.00		$32.00
INSECTICIDE (CUSTOM)	$20.00	1	ACRE	$20.00		$20.00
HERBICIDE (CUSTOM)	$6.00	1	ACRE	$6.00		$6.00
PLASTIC TWINE	$26.00	170	FT/TON	$5.53		$5.53
ESTABLISHMENT: PRINCIPAL		6	YEARS		$35.28	$35.28
INTEREST					$12.64	$12.64
PUMP WATER*		54	AC. IN.	——	——	——
SUBTOTAL				$63.53	$47.93	$111.45

PREHARVEST OPERATIONS	POWER UNIT	ACCOMPLISHMENT RATE		PURCHASED INPUTS	LABOR	FUEL & LUBE	REPAIRS	FIXED COSTS	TOTAL
IRRIGATE (6X)		3.00	HR		$12.00	$94.32	$5.56	$28.94	$140.82
SUBTOTAL		3.00	HR		$12.00	$94.32	$5.56	$28.94	$140.82

HARVEST OPERATIONS	POWER UNIT	ACCOMPLISHMENT RATE		PURCHASED INPUTS	LABOR	FUEL & LUBE	REPAIRS	FIXED COSTS	TOTAL
SWATHER (5X)	14 FT	0.80	HR		$4.00	$3.00	$3.02	$9.47	$19.48
RAKE (5X)	65 HP	0.25	HR		$1.25	$1.34	$0.83	$1.91	$5.33
BALER	130 HP	0.35	HR		$1.75	$2.40	$3.43	$27.03	$34.61
FRONT END LOADER (5X)	96 HP	0.25	HR		$1.25	$1.27	$0.64	$2.11	$5.27
HAUL	SEMI	0.50	HR		$2.50	$2.63	$5.25	$9.79	$20.17
SUBTOTAL		2.15	HR		$10.75	$10.63	$13.16	$50.31	$84.86

OVERHEAD EXPENSES		ACCOMPLISHMENT RATE		PURCHASED INPUTS	LABOR	FUEL & LUBE	REPAIRS	FIXED COSTS	TOTAL
DOWNTIME		0.54	HR		$2.69				$2.69
EMPLOYEE BENEFITS					$3.41				$3.41
INSURANCE				$0.34					$0.34
LAND TAXES								$2.24	$2.24
SUPERVISION AND MANAGEMENT					$25.42				$25.42
OTHER EXPENSES				$17.28					$17.28
SUBTOTAL		0.54	HR	$17.62	$31.52	——	——	$2.24	$51.37
TOTAL OPERATING EXPENSES		5.69	HR	$81.14	$54.27	$104.94	$18.72	$129.42	$388.50
NET OPERATING PROFIT									$46.50
INTEREST ON OPERATING CAPITAL	(58.71 @ 12.00%)								$7.05
INTEREST ON EQUIPMENT INVESTMENT									39.38
RETURN TO LAND AND RISK									$0.07

BUDGET SUMMARY

GROSS RETURN		$435.00	
VARIABLE OPERATING EXPENSES	$204.81		
RETURN OVER VARIABLE EXPENSES		$230.19	(GROSS MARGIN)
FIXED EXPENSES	$129.42		
NET FARM INCOME		$100.77	(RETURN TO CAPITAL, LABOR, LAND & RISK)
LABOR AND MANAGEMENT COSTS	$54.27		
NET OPERATING PROFIT		$46.50	(RETURN TO CAPITAL, LAND & RISK)
CAPITAL COSTS	$46.43		
RETURN TO LAND AND RISK		$0.07	

*Pump water costs are shown under irrigation in the preharvest operations section.

Table 6.10. (continued).

E. Permanent pasture establishment, Class II land, flood irrigation, budgeted per acre costs and returns for Snake Belly Farms, 19X2; planting dates: August 15–October 1

PURCHASED INPUTS	PRICE	QUANTITY		PURCHASED INPUTS	FIXED COSTS	TOTAL
GRASS SEED	$1.00	30	LBS	$30.00		$30.00
NITROGEN (N)	$0.25	16	LBS	$4.00		$4.00
PHOSHATE (P205)	$0.25	20	LBS	$5.00		$5.00
PUMP WATER*		16	AC. IN.			
SUBTOTAL				$39.00		$39.00

PREHARVEST OPERATIONS	POWER UNIT		ACCOMPLISHMENT RATE		PURCHASED INPUTS	LABOR	FUEL & LUBE	REPAIRS	FIXED COSTS	TOTAL
DISC	130	HP	0.17	HR		$0.85	$1.17	$1.84	$4.91	$8.77
PLOW	130	HP	0.48	HR		$2.40	$3.29	$4.44	$10.82	$20.96
DISC (2X)	130	HP	0.34	HR		$1.70	$2.33	$3.67	$9.83	$17.53
PLANE (2X)	96	HP	0.48	HR		$2.40	$2.43	$0.44	$7.21	$12.49
DRILL	63	HP	0.21	HR		$1.05	$1.12	$0.51	$3.93	$6.62
DITCHER (2X)	96	HP	0.10	HR		$0.50	$0.51	$0.08	$1.28	$2.37
IRRIGATE (2X)	96	HP	1.00	HR		$4.00	$27.95	$1.65	$8.58	$42.17
SUBTOTAL			2.78	HR		$12.90	$38.80	$12.63	$46.56	$110.90

OVERHEAD EXPENSES					PURCHASED INPUTS	LABOR	FUEL & LUBE	REPAIRS	FIXED COSTS	TOTAL
DOWNTIME						$2.23				$2.23
EMPLOYEE BENEFITS			0.45	HR		$1.94				$1.94
INSURANCE					$0.19					$0.19
LAND TAXES						$9.15				$9.15
SUPERVISION AND MANAGEMENT										
SUBTOTAL			0.45	HR	$0.19	$13.31				$13.50
TOTAL OPERATING EXPENSES			3.22	HR	$39.19	$26.21	$38.80	$12.63	$46.56	$163.40

*Pump water costs are shown under irrigation in the preharvest operations section.

Table 6.10. (continued).

F. Permanent pasture, Class II land, flood irrigation, budgeted per acre costs and returns for Snake Belly Farms, 19X2; harvesting dates: May 1-November 30

ITEM	PRICE	YIELD	TOTAL
GROSS RETURNS			
ALFALFA HAY	$12.00	14.00 AUMS	$168.00
TOTAL			$168.00

PURCHASED INPUTS	QUANTITY		PRICE	PURCHASED INPUTS	FIXED COSTS	TOTAL
NITROGEN (N)	16	LBS	$0.25	$4.00		$4.00
PHOSPHATE (P205)	20	LBS	$0.25	$5.00		$5.00
LIVESTOCK FAC & EQUIP					$3.00	$3.00
ESTABLISHMENT: PRINCIPAL	15	YEARS			$10.89	$10.89
INTEREST					$9.37	$9.37
PUMP WATER*	36	AC. IN.		$9.00		
SUBTOTAL				$9.00	$23.26	$32.26

PREHARVEST OPERATIONS	POWER UNIT	ACCOMPLISHMENT RATE	PURCHASED INPUTS	LABOR	FUEL & LUBE	REPAIRS	FIXED COSTS	TOTAL
IRRIGATE (6X)		3.00 HR		$12.00	$62.88	$3.71	$19.30	$97.88
SUBTOTAL		3.00 HR		$12.00	$62.88	$3.71	$19.30	$97.88
OVERHEAD EXPENSES								
DOWNTIME				$0.00				$0.00
EMPLOYEE BENEFITS				$1.80				$1.80
INSURANCE			$0.18					$0.18
LAND TAXES							$2.24	$2.24
SUPERVISON AND MANAGEMENT				$11.10				$11.10
OTHER EXPENSES			$17.28					$17.28
SUBTOTAL			$17.46	$12.90			$2.24	$32.60
TOTAL OPERATING EXPENSES		3.00 HR	$26.46	$24.90	$62.88	$3.71	$44.79	$162.73
NET OPERATING PROFIT								$5.27
INTEREST ON OPERATING CAPITAL	(21.15 @ 12.00%)							$2.54
INTEREST ON EQUIPMENT INVESTMENT								$6.27
RETURN TO LAND AND RISK								($3.54)

BUDGET SUMMARY

GROSS RETURN	$168.00	
VARIABLE OPERATING EXPENSES	$93.04	
RETURN OVER VARIABLE EXPENSES	$74.96	(GROSS MARGIN)
FIXED EXPENSES	$44.79	
NET FARM INCOME	$30.17	(RETURN TO CAPITAL, LABOR, LAND & RISK)
LABOR AND MANAGEMENT COSTS	$24.90	
NET OPERATING PROFIT	$5.27	(RETURN TO CAPITAL, LAND & RISK)
CAPITAL COSTS	$8.81	
RETURN TO LAND AND RISK	($3.54)	

*Pump water costs are shown under irrigation in the preharvest operations section.

Table 6.10. (continued).

G. Permanent pasture establishment, Class III land, flood irrigation, budgeted per acre costs and returns for Snake Belly Farms, 19X2; planting dates: August 15-October 1

PURCHASED INPUTS	PRICE	QUANTITY		PURCHASED INPUTS			FIXED COSTS	TOTAL
GRASS SEED	$1.00	30	LBS	$30.00				$30.00
NITROGEN (N)	$0.25	16	LBS	$4.00				$4.00
PHOSHATE (P2O5)	$0.25	20	LBS	$5.00				$5.00
PUMP WATER*		16	AC. IN.					
SUBTOTAL				$39.00				$39.00

PREHARVEST OPERATIONS	POWER UNIT	ACCOMPLISHMENT RATE		PURCHASED INPUTS	LABOR	FUEL & LUBE	REPAIRS	FIXED COSTS	TOTAL
DISC	130 HP	0.17	HR		$0.83	$1.17	$1.84	$4.91	$8.77
PLOW	130 HP	0.48	HR		$2.40	$3.29	$4.44	$10.82	$20.96
DISC (2X)	130 HP	0.34	HR		$1.70	$2.33	$3.67	$9.83	$17.53
PLANE (2X)	96 HP	0.48	HR		$2.40	$2.43	$0.44	$7.21	$12.49
DRILL	63 HP	0.21	HR		$1.05	$1.12	$0.51	$3.99	$6.62
DITCHER (2X)	96 HP	0.10	HR		$0.50	$0.51	$0.08	$1.28	$2.37
IRRIGATE (2X)		1.00	HR		$4.00	$27.95	$1.65	$8.58	$42.17
SUBTOTAL		2.78	HR		$12.90	$38.80	$12.63	$46.56	$110.90

OVERHEAD EXPENSES									
DOWNTIME		0.45	HR		$2.23				$2.23
EMPLOYEE BENEFITS					$1.94				$1.94
INSURANCE				$0.19					$0.19
LAND TAXES								$11.45	$11.45
SUPERVISION AND MANAGEMENT					$9.15				
SUBTOTAL		0.45	HR	$0.19	$13.31	—	—	—	$13.50

| TOTAL OPERATING EXPENSES | | 3.22 | HR | $39.19 | $26.21 | $38.80 | $12.63 | $46.56 | $163.40 |

*Pump water costs are shown under irrigation in the preharvest operations section.

Table 6.10. (continued).

H. Permanent pasture, Class III land, flood irrigation, budgeted per acre costs and returns for Snake Belly Farms, 19X2; harvesting dates: May 1-November 30

ITEM	PRICE	YIELD	PURCHASED INPUTS				FIXED COSTS	TOTAL
GROSS RETURNS								
ALFALFA HAY	$12.00	12.70 AUMS						$152.40
TOTAL								$152.40

PURCHASED INPUTS		PRICE	QUANTITY		PURCHASED INPUTS		FIXED COSTS	TOTAL
NITROGEN (N)		$0.25	16	LBS	$4.00			$4.00
PHOSPHATE (P205)		$0.25	20	LBS	$5.00			$5.00
LIVESTOCK FAC & EQUIP							$3.00	$3.00
ESTABLISHMENT: PRINCIPAL			15	YEARS			$10.89	$10.89
INTEREST							$9.37	$9.37
PUMP WATER*			36	AC. IN.	$9.00			
SUBTOTAL					$9.00		$23.26	$32.26

PREHARVEST OPERATIONS	POWER UNIT	ACCOMPLISHMENT RATE		PURCHASED INPUTS	LABOR	FUEL & LUBE	REPAIRS	FIXED COSTS	TOTAL
IRRIGATE (6X)		3.00	HR		$12.00	$62.88	$3.71	$19.30	$97.88
SUBTOTAL		3.00	HR		$12.00	$62.88	$3.71	$19.30	$97.88

OVERHEAD EXPENSES	PURCHASED INPUTS	LABOR	FIXED COSTS	TOTAL
DOWNTIME		$0.00		$0.00
EMPLOYEE BENEFITS		$1.80		$1.80
INSURANCE	$0.18			$0.18
LAND TAXES			$2.24	$2.24
SUPERVISION AND MANAGEMENT		$10.52		$10.52
OTHER EXPENSES	$17.28			$17.28
SUBTOTAL	$17.46	$12.12	$2.24	$31.82
TOTAL OPERATING EXPENSES	$26.46	$24.12	$62.88 ($3.71)	$44.79	$161.95

NET OPERATING PROFIT	($9.55)
INTEREST ON OPERATING CAPITAL (21.15 @ 12.00%)	$2.54
INTEREST ON EQUIPMENT INVESTMENT	$6.27
RETURN TO LAND AND RISK	($18.36)

BUDGET SUMMARY

GROSS RETURN	$152.40	
VARIABLE OPERATING EXPENSES	$93.04	
RETURN OVER VARIABLE EXPENSES	$59.36	(GROSS MARGIN)
FIXED EXPENSES	$44.79	
NET FARM INCOME	$14.57	(RETURN TO CAPITAL, LABOR, LAND & RISK)
LABOR AND MANAGEMENT COSTS	$24.12	
NET OPERATING PROFIT	$9.55	(RETURN TO CAPITAL, LAND & RISK)
CAPITAL COSTS	$8.81	
RETURN TO LAND AND RISK	($18.36)	

*Pump water costs are shown under irrigation in the preharvest operations section.

Table 6.10. (continued).

I. Wheat, flood irrigation, budgeted per acre costs and returns for Snake Belly Farms, 19X2; planting dates: August 15-October 1; harvesting dates: May 20-October 15

ITEM	PRICE	YIELD		TOTAL
GROSS RETURNS				
WHEAT	$2.30	70.00 BU		$161.00
ASCS DEFICIENCY	$2.00	70.00 BU		$140.00
ASCS DIVERSION	$0.00			$0.00
GRAZING	$44.00	3.80 AUM		$167.20
TOTAL				$468.20

PURCHASED INPUTS	PRICE	QUANTITY	PURCHASED INPUTS	TOTAL
SEED	$0.20	100 LBS	$20.00	$20.00
NITROGEN (N)	$0.25	150 LBS	$37.50	$37.50
PHOSPHATE (P205)	$0.25	50 LBS	$12.50	$12.50
INSECTICIDE	$12.00	1 ACRE	$12.00	$12.00
PUMP WATER*		28 AC. IN.	——	——
SUBTOTAL			$82.00	$82.00

PREHARVEST OPERATIONS	POWER UNIT	ACCOMPLISHMENT RATE	PURCHASED INPUTS	LABOR	FUEL & LUBE	REPAIRS	FIXED COSTS	TOTAL
PLOW	130 HP	0.48 HR		$2.40	$3.29	$4.44	$10.82	$20.96
DISC (2X)	130 HP	0.34 HR		$1.70	$2.33	$3.67	$9.83	$17.53
FERTILIZE	DEALER APPLIED							
DRILL	65 HP	0.21 HR		$1.05	$1.12	$0.51	$3.93	$6.62
DITCHER (2X)	96 HP	0.10 HR		$0.50	$0.51	$0.08	$1.28	$2.37
IRRIGATE (4X)		2.00 HR		$8.00	$48.90	$2.88	$15.01	$74.79
SUBTOTAL		3.13 HR		$13.65	$56.16	$11.59	$40.97	$122.27

HARVEST OPERATIONS

			PURCHASED INPUTS					TOTAL
COMBINE (CUSTOM)			$18.00					$18.00
HAUL (CUSTOM)			$8.40					$8.40
SUBTOTAL			$26.40					$26.40

OVERHEAD EXPENSES

		ACCOMPLISHMENT RATE	PURCHASED INPUTS	LABOR			FIXED COSTS	TOTAL
DOWNTIME		0.28 HR		$1.41				$1.41
EMPLOYEE BENEFITS				$2.05				$2.05
INSURANCE			$0.20					$0.20
LAND TAXES							$2.24	$2.24
SUPERVISION AND MANAGEMENT				$25.67				$25.67
OTHER EXPENSES			$17.28					$17.28
SUBTOTAL		0.28 HR	$17.48	$29.13			$2.24	$48.85
TOTAL OPERATING EXPENSES		3.41 HR	$125.88	$42.78	$56.16	$11.59	$43.11	$279.53

NET OPERATING PROFIT				$188.67
INTEREST ON OPERATING CAPITAL	(60.14 @ 12.00%)			$7.22
INTEREST ON EQUIPMENT INVESTMENT				$22.06
RETURN TO LAND AND RISK				$159.40

BUDGET SUMMARY

GROSS RETURN		$468.20	
VARIABLE OPERATING EXPENSES	$193.63		
RETURN OVER VARIABLE EXPENSES		$274.57	(GROSS MARGIN)
FIXED EXPENSES	$43.11		
NET FARM INCOME		$231.46	(RETURN TO CAPITAL, LABOR, LAND & RISK)
LABOR AND MANAGEMENT COSTS	$42.78		
NET OPERATING PROFIT		$188.67	(RETURN TO CAPITAL, LAND & RISK)
CAPITAL COSTS	$29.28		
RETURN TO LAND AND RISK		$159.40	

*Pump water costs are shown under irrigation in the preharvest operations section.

Table 6.10. (continued).

J. Soybeans, flood irrigation, budgeted per acre costs and returns for Snake Belly Farms, 19X2; planting dates: April 15-June 1; harvesting dates: September 15-October 15

ITEM	PRICE	YIELD	TOTAL
GROSS RETURNS			
SOYBEANS	$5.40	45.00 BU	$243.00
TOTAL			$243.00

PURCHASED INPUTS	PRICE	QUANTITY		PURCHASED INPUTS	TOTAL
SEED	$0.40	20	LBS	$8.00	$8.00
NITROGEN (N)	$0.25	32	LBS	$8.00	$8.00
PHOSPHATE (P205)	$0.25	40	LBS	$10.00	$10.00
HERBICIDE	$9.00	1	X/ACRE	$9.00	$9.00
HERBICIDE (CUSTOM)	$10.00	1	X/ACRE	$10.00	$10.00
INSECTICIDE (CUSTOM)	$10.00	1	X/ACRE	$10.00	$10.00
CROP INSURANCE	$400	/ACRE		$20.00	$20.00
PUMP WATER*		21	AC. IN.	——	——
SUBTOTAL				$75.00	$75.00

PREHARVEST OPERATIONS	POWER UNIT	ACCOMPLISHMENT RATE		PURCHASED INPUTS	LABOR	FUEL & LUBE	REPAIRS	FIXED COSTS	TOTAL
DISC	130 HP	0.17	HR		$0.85	$1.17	$1.84	$4.91	$8.77
PLOW	130 HP	0.48	HR		$2.40	$3.29	$4.44	$10.82	$20.96
FLOAT	96 HP	0.16	HR		$0.80	$0.81	$0.20	$1.48	$3.30
FERTILIZE	DEALER APPLIED								
LISTER	96 HP	0.18	HR		$0.90	$0.91	$0.49	$3.74	$6.04
PRE-IRRIGATE		0.75	HR		$3.00	$9.71	$0.36	$1.88	$14.41
CULT & SPRAY	65 HP	0.26	HR		$1.30	$1.39	$0.77	$3.85	$7.31
PLANTER	96 HP	0.26	HR		$1.30	$1.32	$0.61	$10.49	$13.72
CULTIVATOR (3X)	65 HP	0.63	HR		$3.15	$3.37	$1.21	$3.01	$10.74
DITCHER (2X)	96 HP	0.10	HR		$0.50	$0.51	$0.08	$1.28	$2.37
IRRIGATE (3X)		1.50	HR		$6.00	$27.51	$1.62	$8.44	$43.57
					——	——	——	——	——
SUBTOTAL		4.49	HR		$20.20	$49.45	$11.63	$49.90	$131.17

HARVEST OPERATIONS			PURCHASED INPUTS					TOTAL
COMBINE (CUSTOM)			$18.00					$18.00
HAUL (CUSTOM)			$5.40					$5.40
			——					——
SUBTOTAL			$23.40					$23.40

OVERHEAD EXPENSES			PURCHASED INPUTS	LABOR	FUEL & LUBE	REPAIRS	FIXED COSTS	TOTAL
DOWNTIME	0.19	HR		$0.94				$0.94
EMPLOYEE BENEFITS				$3.03				$3.03
INSURANCE			$0.30					$0.30
LAND TAXES							$2.24	$2.24
SUPERVISION AND MANAGEMENT				$13.79				$13.79
OTHER EXPENSES			$17.28					$17.28
			——	——			——	——
SUBTOTAL	0.19	HR	$17.58	$17.76	——	——	$2.24	$37.58

TOTAL OPERATING EXPENSES	4.68	HR	$115.98	$37.96	$49.45	$11.63	$52.14	$267.15

NET OPERATING PROFIT			($24.15)
INTEREST ON OPERATING CAPITAL	(52.77 @ 12.00%)		$6.33
INTEREST ON EQUIPMENT INVESTMENT			$31.28
RETURN TO LAND AND RISK			($61.77)

BUDGET SUMMARY

GROSS RETURN		$243.00	
VARIABLE OPERATING EXPENSES	$177.06		
RETURN OVER VARIABLE EXPENSES		$65.94	(GROSS MARGIN)
FIXED EXPENSES	$52.14		
NET FARM INCOME		$13.81	(RETURN TO CAPITAL, LABOR, LAND & RISK)
LABOR AND MANAGEMENT COSTS	$37.96		
NET OPERATING PROFIT		($24.15)	(RETURN TO CAPITAL, LAND & RISK)
CAPITAL COSTS	$37.62		
RETURN TO LAND AND RISK		($61.77)	

*Pump water costs are shown under irrigation in the preharvest operations section.

110

Table 6.10. (continued).

K. Upland cotton (picker), flood irrigation, budgeted per acre costs and returns for Snake Belly Farms, 19X2; planting dates: April 15-June 1; harvesting dates: October 1-December 31

ITEM	PRICE	YIELD		TOTAL
GROSS RETURNS				
COTTON LINT	$0.60	750	LBS	$450.00
COTTON SEED	$0.03	1200	LBS	$36.00
ASCS DEFICIENCY	$0.2715	750	LBS	$203.63
ASCS DIVERSION	$0.00			$0.00
TOTAL				$689.63

PURCHASED INPUTS	PRICE	QUANTITY		PURCHASED INPUTS	FIXED COSTS	TOTAL
SEED	$0.40	20	LBS	$8.00		$8.00
NITROGEN (N)	$0.25	32	LBS	$8.00		$8.00
PHOSPHATE (P205)	$0.25	40	LBS	$10.00		$10.00
HERBICIDE	$9.00	1	ACRE	$9.00		$9.00
HERBICIDE (CUSTOM)	$10.00	1	ACRE	$10.00		$10.00
INSECTICIDE (CUSTOM)	$25.00	1	/ACRE	$25.00		$25.00
CROP INSURANCE		400	DOLLARS	$20.00		$20.00
PUMP WATER*		28	AC. IN.	——		——
SUBTOTAL				$75.00		$90.00

PREHARVEST OPERATIONS	POWER UNIT	ACCOMPLISHMENT RATE		PURCHASED INPUTS	LABOR	FUEL & LUBE	REPAIRS	FIXED COSTS	TOTAL
DISC	130 HP	0.17	HR		$0.85	$1.17	$1.84	$4.91	$8.77
PLOW	130 HP	0.48	HR		$2.40	$3.29	$4.44	$10.82	$20.96
FLOAT	96 HP	0.16	HR		$0.80	$0.81	$0.20	$1.48	$3.30
FERTILIZE	DEALER APPLIED								
LISTER	96 HP	0.18	HR		$0.90	$0.91	$0.49	$3.74	$6.04
PRE-IRRIGATE		0.75	HR		$3.00	$12.23	$0.72	$3.75	$19.70
CULT & SPRAY	65 HP	0.26	HR		$1.30	$1.39	$0.77	$3.85	$7.31
PLANTER	96 HP	0.26	HR		$1.30	$1.32	$0.61	$10.49	$13.72
CULTIVATOR (3X)	65 HP	0.63	HR		$3.15	$3.37	$1.21	$3.01	$10.74
HAND HOE (CUSTOM)				$20.00					$20.00
DITCHER (2X)	96 HP	0.10	HR		$0.50	$0.51	$0.08	$1.28	$2.37
IRRIGATE (3X)		1.50	HR		$6.00	$36.68	$2.16	$11.2644	$56.10
SUBTOTAL		4.49	HR	$20.00	$20.20	$61.68	$12.53	$54.59	$168.99

HARVEST OPERATIONS									
COTTON PICKER	2 ROW	1.24	HR		$6.20	$7.50	$24.82	$55.93	$94.45
HAUL (2X)	65 HP	1.00	HR		$5.00	$5.35	$0.91	$8.41	$19.67
GIN COTTON (CUSTOM)				$84.75					$84.75
SUBTOTAL		2.24	HR	$84.75	$11.20	$12.85	$25.73	$64.34	$198.87

POSTHARVEST OPERATIONS									
SHREDDER	96 HP	0.14	HR		$0.70	$0.71	$0.16	$5.29	$6.86
SUBTOTAL		0.14	HR		$0.70	$0.71	$0.16	$5.29	$6.86

OVERHEAD EXPENSES									
DOWNTIME		1.16	HR		$5.78				$5.78
EMPLOYEE BENEFITS					$4.81				$4.81
INSURANCE				$0.48					$0.48
LAND TAXES								$2.24	$2.24
SUPERVISION AND MANAGEMENT					$38.54				$38.54
OTHER EXPENSES				$17.28					$17.28
SUBTOTAL		1.16	HR	$17.76	$49.13	——	——	$2.24	$69.13

TOTAL OPERATING EXPENSES		8.03	HR	$212.51	$81.23	$75.23	$38.42	$126.45	$533.85
NET OPERATING PROFIT									$155.78
INTEREST ON OPERATING CAPITAL	(66.84 @ 12.00%)								$8.02
INTEREST ON EQUIPMENT INVESTMENT									$31.28
RETURN TO LAND AND RISK									$116.48

BUDGET SUMMARY

GROSS RETURN		$689.63	
VARIABLE OPERATING EXPENSES	$326.16		
RETURN OVER VARIABLE EXPENSES		$363.46	(GROSS MARGIN)
FIXED EXPENSES	$126.45		
NET FARM INCOME		$237.01	(RETURN TO CAPITAL, LABOR, LAND & RISK)
LABOR AND MANAGEMENT COSTS	$81.23		
NET OPERATING PROFIT		$155.78	(RETURN TO CAPITAL, LAND & RISK)
CAPITAL COSTS	$39.30		
RETURN TO LAND AND RISK		$116.48	

*Pump water costs are shown under irrigation in the preharvest operations section.

Table 6.10. (*continued*).

L. Corn for grain, flood irrigation, budgeted per acre costs and returns for Snake Belly Farms, 19X2; planting dates: April 15-May 1; harvesting dates: October 1-November 1

ITEM	PRICE	YIELD	TOTAL
GROSS RETURNS			
CORN FOR GRAIN	$2.10	170 BU	$357.00
ASCS DEFICIENCY	$1.21	170 BU	$205.70
ASCS DIVERSION	$2.00	24 BU	$48.00
TOTAL			$610.70

PURCHASED INPUTS	PRICE	QUANTITY		PURCHASED INPUTS			FIXED COSTS	TOTAL
SEED	$0.75	32	000	$24.00				$24.00
NITROGEN (N)	$0.25	200	LBS	$50.00				$50.00
INSECTICIDE (CUSTOM)(2X)	$25.00	1	ACRE	$25.00				$25.00
CROP INSURANCE		400	DOLLARS	$20.00				$20.00
PUMP WATER*		48	AC. IN.	——				——
SUBTOTAL				$119.00				$119.00

PREHARVEST OPERATIONS	POWER UNIT	ACCOMPLISHMENT RATE		PURCHASED INPUTS	LABOR	FUEL & LUBE	REPAIRS	FIXED COSTS	TOTAL
DISC	130 HP	0.17	HR		$0.85	$1.17	$1.84	$4.91	$8.77
PLOW	130 HP	0.48	HR		$2.40	$3.29	$4.44	$10.82	$20.96
FLOAT	96 HP	0.16	HR		$0.80	$0.81	$0.20	$1.48	$3.30
FERTILIZE	DEALER APPLIED								
LISTER	96 HP	0.18	HR		$0.90	$0.91	$0.49	$3.74	$6.04
PRE-IRRIGATE		0.75	HR		$3.00	$13.97	$0.82	$4.29	$22.08
CULT & SPRAY	65 HP	0.26	HR		$1.30	$1.39	$0.77	$3.85	$7.31
PLANTER	96 HP	0.26	HR		$1.30	$1.32	$0.61	$10.49	$13.72
CULTIVATOR (2X)	65 HP	0.42	HR		$2.10	$2.25	$0.81	$2.01	$7.16
DITCHER (2X)	96 HP	0.10	HR		$0.50	$0.51	$0.08	$1.28	$2.37
IRRIGATE (5X)		2.50	HR		$10.00	$69.86	$4.12	$21.44	$105.42
SUBTOTAL		5.28	HR		$23.15	$95.48	$14.18	$64.31	$197.12

HARVEST OPERATIONS				PURCHASED INPUTS					TOTAL
COMBINE (CUSTOM)				$37.40					$37.40
HAUL (CUSTOM)				$18.70					$18.70
SUBTOTAL				$56.10					$56.10

OVERHEAD EXPENSES				PURCHASED INPUTS	LABOR			FIXED COSTS	TOTAL
DOWNTIME		0.51	HR		$2.54				$2.54
EMPLOYEE BENEFITS					$3.47				$3.47
INSURANCE				$0.35					$0.35
LAND TAXES								$2.24	$2.24
SUPERVISION AND MANAGEMENT					$34.33				$34.33
OTHER EXPENSES				$17.28					$17.28
SUBTOTAL		0.51	HR	$17.62	$40.34			$2.24	$60.20

TOTAL OPERATING EXPENSES		5.79	HR	$192.72	$63.49	$95.48	$14.18	$66.54	$432.42

NET OPERATING PROFIT									$178.28
INTEREST ON OPERATING CAPITAL	(86.92 @ 12.00%)								$10.43
INTEREST ON EQUIPMENT INVESTMENT									$35.45
RETURN TO LAND AND RISK									$132.40

BUDGET SUMMARY

GROSS RETURN		$610.70	
VARIABLE OPERATING EXPENSES	$302.39		
RETURN OVER VARIABLE EXPENSES		$308.31	(GROSS MARGIN)
FIXED EXPENSES	$66.54		
NET FARM INCOME		$241.76	(RETURN TO CAPITAL, LABOR, LAND & RISK)
LABOR AND MANAGEMENT COSTS	$63.49		
NET OPERATING PROFIT		$178.28	(RETURN TO CAPITAL, LAND & RISK)
CAPITAL COSTS	$45.88		
RETURN TO LAND AND RISK		$132.40	

*Pump water costs are shown under irrigation in the preharvest operations section.

Table 6.10. (continued).

M. Summary of per acre costs and returns for Snake Belly Farms, 19X2

	ALFALFA EST I	ALFALFA HAY!	ALFALFA EST II	ALFALFA HAY II	PASTURE EST II	PASTURE II	PASTURE EST III	PASTURE III	WHEAT	SOYBEANS	PICKER COTTON	CORN
YIELD		7.00		5.00		14.00		12.70	70.00 / 3.80	45.00	750.00 / 1200.00	170.00
GROSS RETURN		$609.00		$435.00		$168.00		$152.40	$468.20	$243.00	$689.63	$610.70
CASH OPERATING EXPENSES												
SEED	$75.00		$75.00									
FERTILIZER		$40.00		$32.00	$30.00		$30.00		$20.00	$8.00	$8.00	$24.00
CHEMICALS	$10.00	$26.00	$10.00	$26.00					$50.00	$18.00	$18.00	$50.00
CROP INSURANCE					$9.00	$9.00	$9.00	$9.00	$12.00	$29.00	$44.00	$25.00
OTHER PURCHASED INPUTS										$20.00	$20.00	$25.00
CANAL WATER		$7.74		$5.53								
FUEL, OIL, LUBRICANT - EQUIP	$10.86	$10.63	$10.86	$10.63	$10.86	$62.88	$10.86	$62.88	$7.26	$12.77	$26.33	$25.62
FUEL IRRIGATION	$27.95	$94.32	$27.95	$94.32	$27.95	$3.71	$27.95	$3.71	$48.90	$36.68	$48.90	$69.86
REPAIRS	$12.63	$18.72	$12.63	$18.72	$12.63		$12.63		$11.59	$11.59	$38.42	$14.18
CUSTOM CHARGES									$26.40	$26.40	$104.75	$56.10
LAND TAXES	$0.19	$2.24	$0.19	$2.24	$0.19	$2.24	$0.19	$2.24	$2.24	$2.24	$2.24	$2.24
OTHER EXPENSES		$17.62		$17.62		$17.46		$17.46	$17.48	$17.48	$17.76	$19.87
TOTAL CASH EXPENSES	$136.63	$217.26	$136.63	$207.05	$90.63	$95.28	$90.63	$95.28	$195.87	$179.29	$328.40	$304.63
RETURN OVER CASH EXPENSES	($136.63)	$391.74	($136.63)	$227.95	($90.63)	$72.72	($90.63)	$57.12	$272.33	$63.71	$361.23	$306.07
FIXED EXPENSES	$46.56	$127.18	$46.56	$127.18	$46.56	$42.55	$46.56	$42.55	$40.87	$49.90	$124.22	$64.30
TOTAL EXPENSES	$183.19	$344.44	$183.19	$334.23	$137.19	$137.83	$137.19	$137.83	$236.74	$229.19	$452.62	$368.93
NET FARM INCOME	($183.19)	$264.56	($183.19)	$100.77	($137.19)	$30.17	($137.19)	$14.57	$231.46	$13.81	$237.01	$241.77
LABOR AND MANAGEMENT COSTS	$28.51	$62.97	28.51	$54.27	$26.21	$24.90	$26.21	$24.12	$42.78	$37.96	$81.23	$63.49
NET OPERATING PROFIT	($211.70)	$201.59	($211.70)	$46.50	($163.40)	$5.27	($163.40)	($9.55)	$188.67	($24.15)	$155.78	$178.28
CAPITAL COSTS												
INTEREST ON OPERATING CAPITAL		$7.66		$7.05		$2.54		$2.54	$7.22	$6.33	$8.02	$10.43
INTEREST ON EQUIPMENT INVESTMENT		$39.38		$39.38		$6.27		$6.27	$22.06	$31.28	$31.28	$35.45
TOTAL CAPITAL COSTS	$0.00	$47.04	$0.00	$46.43	$0.00	$8.81	$0.00	$8.81	$29.28	$37.62	$39.30	$45.88
RETURN TO LAND AND RISK	($211.70)	$154.55	($211.70)	$0.07	($163.40)	($3.54)	($163.40)	($18.36)	$159.40	($61.77)	$116.48	$132.40

CHAPTER 7

Purchasing Schedules and Inventory Planning

Inventories of stored crops, stored feed, livestock on feed, and inputs on hand can represent a large drain on the cash flow and cash reserves of a farm or ranch, because they tie up funds that could be used for other productive or investment purposes. The output and input marketing principles discussed in Chapter 5 dominate many of the inventory purchasing and control decisions on a typical farm or ranch. Safety stock concerns are especially important in determining the manner in which feed and other input inventories must be purchased or transferred from inventories of crops available for sale. In some respects, these are the same concerns that manufacturing firms might have.[1] However, the similarities between farms and ranches on one hand and manufacturing or retailing firms on the other soon end.

The fluidity of most agricultural markets makes inventory planning, input purchasing, and inventory control decisions much different than those of the manufacturing or retailing firm; feed can be bought and sold quickly, and most crop and livestock commodities can be sold within a few hours. Farms and ranches, further, seldom need safety stocks of fertilizers, but may have need for quantities of feed to get through a week-long storm, for enough baling wire to at least cover the next hay cut, or for enough insecticides to allow a quick response to a common problem.

Much of the recent inventory control literature has focused on a concept first adopted widely by Japanese manufacturing firms—the Just-in-Time inventory control system, which minimizes inventory costs through deliveries of inputs just in time to be used in the manufacturing process. In other words, there is no inventory of inputs; the inputs arrive at the factory when needed, not days, weeks, or months before. There is likely not a great deal of direct applicability of a just-in-time system to agricultural production firms, except of course that by thinking about this process we understand that inventory carrying costs are controllable. One possible application is to livestock feeding operations that might purchase grains and/or some forages for delivery the day they are needed (assuming reasonable safety stock levels are maintained to anticipate a freak snowstorm or the like). But many agricultural inputs do not meet the flow characteristics of a normal just-in-time system. Generally, fertilizer is applied at just a few points in time during the growing cycle, and forages are seldom ensiled for later sale and delivery to the feeder. Again, while we can learn from just-in-time and economic order quantity models, direct applications are probably few in production agriculture.

The inventory of produced agricultural commodities is also a much different problem than that of manufacturing or retailing firms, because we can sell quickly. The problem, or at least the reason why the commodities are being stored in the first place, is that although we can sell quickly, we may not be able to obtain the desired price. We must recognize seasonal price patterns and reconcile our needs for immediate cash and our desire to maximize net price. But first we must deal with a more mundane problem of simply tracking inventories and projecting purchases of inputs and the amounts of outputs and inputs available for feed, sale, application, or other uses. In a later section of this chapter we will address inventory carrying costs, but we will hold discussions of revisions of production, inventory, and marketing plans to optimize our final cash flow plan until Chapter 9.

SOURCES AND USES

The tracking of physical units of all inventories begins with an understanding of sources and uses concepts, a simplified flow plan. No matter what the item, whether it is corn, dollars, cows, or nails, when we consider any production period, the inventory flow pattern will always adhere to one simple equation: Sources = Uses.

For any particular commodity, the sources must equal the uses; nothing vanishes into thin air or is created by magic. For livestock, the Sources = Uses equation can be expanded into:

      Beginning inventory
+ Purchases
+ Number of animals weaned (or transferred from another class)
− Death loss
− Personal consumption
− Sales (or transferred to another class)
= Ending inventory

For crops and feed, a similar expansion, obviously without the death loss, is possible.

      Beginning inventory
+ Production
+ Purchases
− Uses for feed or seed
− Sales
− Losses (due to drying, handling, or pests)
= Ending inventory

And finally, a similar argument could be made for baling wire (or any other purchased input for that matter):

Beginning inventory
+ Purchases
− Uses for baling
− Losses[2]
= Ending inventory

These equations are not specific with respect to time period. We usually think of these equations as applying to a full production year, but it may be more useful to set up a system whereby we can track each major inventory item on a more frequent basis, such as monthly, so that we can match the periods within our cash flow budget.

LIVESTOCK NUMBERS: SOURCES AND USES

The Agricultural Financial Reporting and Analysis (AFRA) system offers an excellent schedule for tracking livestock. If we are dealing with a beef cow herd, it would usually be adequate to consider the annual flow of animals. Under the feeder stock category (the top half of Schedule C-4 shown in Figure 7.1), we should list the number of calves or light yearlings that were held over from last year's weanings. The type of animal should be specified, i.e., 400–500-pound heifer calves or 500–700-pound light yearling steers. Any animals in these categories that are purchased or born during the year should also be listed, all in the left half of Schedule C-4. These three divisions, beginning inventory, purchases, and numbers born (or weaned) represent the total sources of animals. The right half shows the uses of the animals: the number that died, the projection of sales, and finally, the ending inventory. One use is missing, however. The missing category is transfer from one class to another. For example, some or all of the yearling heifers might be designated as cow herd replacements and held over beyond the normal sale time. If we make the argument that the uses side lacks a column to account for out-transfers, we must then also argue that the left side lacks a column for in-transfers, a source of animals when disaggregating by type or kind. For some farms and ranches, it may make more sense to dedicate a separate page of Schedule C-4 to each type of livestock, such as separate pages for feeder steers, feeder heifers, and swine. Using a separate page would allow a monthly track of inventory.

An excellent alternative to AFRA Schedule C-4 for swine is shown in Figure 7.2. The Illinois Sow-Month Breeding Herd Record accomplishes the same thing as AFRA Schedule C-4 and more; it directs breeding herd efficiency analysis as well. Regardless of whether C-4, the Illinois hog form, or a similar self-developed form for another type of livestock is used, the important element here is that for a type of livestock with frequent herd inventory changes it will be necessary to track inventory on a more frequent basis than an annual projection. At the very minimum, the items that use or produce cash, purchases, and sales must be projected on a monthly basis in order to provide the information required for the cash flow budget.

| | | C-4 | | BUDGET SHEET FOR LIVESTOCK PRODUCTION | | | | | | | | | | | | | | For 19____ C-4 |

Kind	Beginning Inventory			Purchases					No. Raised	Transferred in (out)	No. Died	Sales					Ending Inventory		
	No.	Wt./Hd.	Value	Mo.	No.	Wt./Hd.	$/Unit	Cost				Mo.	No.	Wt./Hd.	$/Unit	Value	No.	Wt./Hd.	Value
FEEDER STOCK — LIVESTOCK AND POULTRY																			
			$				$	$							$	$			$
BREEDING STOCK — LIVESTOCK AND POULTRY																			
			$				$	$							$	$			$

This form is copyrighted. It is a violation of the U.S. Copyright Law to reproduce it in any manner. To order forms write or call Century Communications Inc., 6201 Howard St., Niles, IL 60714; 708/647-1200 or Deere Agricultural Services Co., 11701 Borman Dr., St. Louis, MO 63146; 314/569-2700.

Figure 7.1. Budget sheet for livestock production, AFRA Schedule C-4.

Source: A.W. Oltmans, D.A. Klinefelter, and T.L. Frey, *Agricultural Financial Reporting and Analysis,* Century Communications, Inc, Niles, Ill., 1992.

CROP AND FEED SOURCES AND USES

The AFRA Budget Sheet for Crop Usage and Feed Requirements (Schedule C-5, Figure 7.3) works in much the same way as the livestock schedule, or at least the top half does. When projecting inventory sources and uses for crops and feed, we clearly do not need to be concerned with births and deaths, but the sources are surprisingly similar: beginning inventory, production, and purchases. Those sources (again on the left side) must equal the uses on the right side: feed use, on-farm seed use, sales, and ending inventory. The uses side, however, should also account for losses due to shrinkage, rodents, and waste.

For multiple-harvest crops like alfalfa, it would probably be useful in scheduling to use a separate C-5 page and track inventory changes on a monthly basis. A similar monthly track might also be useful for crops like corn or grain sorghum that will be fed to on-farm livestock.

In addition to physical amounts, both budget sheets C-4 and C-5 provide locations for inventory values as well as values for cash transactions. This valuation of inven-

All Sows

All Gilts

Combined

FEMALE MONTH BREEDING HERD RECORD

name _____   county _____

Changes during the Month of	Number of Females		Total Females in Breeding Herd at End of Month (Female Months)[2/]		Number of Pigs:	
	Added[1/]	Sold, Died & Transferred			Farrowed during month	Weaned during month
			(Jan. 1)	( )		
January			Jan. 31			
February			Feb. 28			
March			Mar. 31			
April			Apr. 30			
May			May 31			
June			June 30			
July			July 31			
August			Aug. 31			
September			Sept. 30			
October			Oct. 31			
November			Nov. 30			
December			Dec. 31			
TOTALS						3/
FEMALE YEARS (Divide Total Female Months by 12)						
PIGS FARROWED PER FEMALE YEAR						
PIGS WEANED PER FEMALE YEAR						

1/ Gilts are counted as added to the breeding herd in the month they reach market weight.

2/ Previous month's total plus number added minus number sold, died, and transferred equals total at end of current month.

Number of unweaned pigs on inventory, January 1. _____ 3/
Number of unweaned pigs on inventory, December 31. _____ 3/

3/ The total number of pigs weaned for the year should be adjusted by the difference in number of unweaned pigs on inventory, January 1 and December 31.

A copy of the female month breeding herd record can be forwarded to the Urbana office for summarization at check-in time by the fieldman.

Figure 7.2. Illinois sow-month breeding herd record.

Note: Adopted for use with the Illinois Farm Business Record in cooperation with the Illinois FBFM Association and the Department of Agricultural Economics, University of Illinois, Urbana.

tories, though not necessary for cash flow purposes, can be done easily while thinking about commodity flows and marketing plans and will be an excellent reference when completing projected balance sheets and income statements.

The bottom half of the budget sheet for crop usage and feed requirements should be used by livestock producers. Planned numbers of livestock and their feed requirements (from old crop or beginning inventory or from this year's production) should be listed so that the feed column at the top half of the page will be the best estimates of feed use.

BUDGET SHEET FOR CROP USAGE AND FEED REQUIREMENTS   For 19____

C-5

Crop	Beginning Inventory		Produced	Purchases				Farm Use		Sales				Ending Inventory	
	Quantity	Value		Mo.	Quantity	$/Unit	Cost	Feed	Seed	Mo.	Quantity	$/Unit	Value	Quantity	Value
		$	//////			$	$					$	$	////////	//////////
	//////	//////												////////	////////// $
			//////											////////	//////////
	//////	//////												////////	//////////
			//////											////////	//////////
	//////	//////												////////	//////////
			//////											////////	//////////
	//////	//////												////////	//////////
			//////											////////	//////////
	//////	//////												////////	//////////
			//////											////////	//////////
	//////	//////												////////	//////////

LIVESTOCK FEED REQUIREMENTS

Livestock		Period On Feed	Corn, bu.				Supplement	Hay, tons		Silage, tons	
Kind	No.		Old	New	Old	New	Tons	Old	New	Old	New
TOTALS											

This form is copyrighted. It is a violation of the U.S. Copyright Law to reproduce it in any manner. To order forms write or call Century Communications Inc., 6201 Howard St., Niles, IL 60714; 708/647-1200 or Doane Agricultural Services Co., 11701 Borman Dr., St. Louis, MO 63146; 314/569-2700.

Figure 7.3. Budget sheet for crop usage and feed requirements, AFRA Schedule C-5.

Source: A.W. Oltmans, D.A. Klinefelter, and T.L. Frey, *Agricultural Financial Reporting and Analysis,* Century Communications, Inc, Niles, Ill., 1992.

	Beginning Inventory		Purchases						Ending Inventory	
Item	Quantity	Value	Month	Quantity	$/Unit	Cost	Losses	Uses	Quantity	Value

Figure 7.4. Budget sheet for purchased inputs.

INPUT SOURCES AND USES

A simpler schedule (like the example shown in Figure 7.4) can be used for purchased inputs not produced on the farm or ranch. Similar requirements that sources must equal uses remain in force. Depending upon the frequency of use, an annual or monthly track can be maintained.

By completing the enterprise budgets and the whole-farm budget in Chapter 6, two of the beginning, critical schedules of the AFRA cash flow budgeting process have been nearly completed in a different form. That information should be rewritten onto the AFRA forms in order to maintain a consistent, complete package. The main advantage of completing the Budget Sheet for Crop Production (Schedule C-2, Figure 7.5) is the multiplication of seed, fertilizer, and chemical needs on a per-acre basis by the number of acres to be planted that results in total needs for a particular seed variety, fertilizer formulation, or chemical for the entire crop.

Total Seed, Fertilizer, and Chemicals

The total crop-by-crop needs should be carried to the top half of the Summary of Crop Costs (Schedule C-3, Figure 7.6). The need for each particular seed variety should be listed and the total needs for each particular fertilizer formulation and chemical should be totaled across crops to compute total farm needs for each item. Once the total farm needs for seed, fertilizer, and chemicals are known, we can refer to the input marketing techniques discussed in Chapter 5 to price and schedule the purchases of these inputs. Schedule C-3 requires that the type of input be listed along with total farm needs, price per unit (the multiplication of these two numbers yields total cost), and the month of purchase. This final result produces a summary of costs for transfer to the cash flow budget and a purchasing schedule that can be used to begin the ordering-contracting process for acquisition of the inputs.

Fuel, Custom Hire, and Labor

While we have not stressed the purchasing or use of fuel, custom hire, and labor hiring needs in this chapter so far, the purchasing needs for each of these items can be determined from the enterprise budgets. Custom hiring needs come directly from the enterprise budgets and can be summarized on the bottom half of Schedule C-3.

The estimation of labor needs is a bit more difficult in that labor requirements should be summarized from the budgets in at least three categories:

1. Operator labor (unpaid)
2. Full-time hired labor
3. Seasonal labor

A fourth category, unpaid family labor, should also be listed if it is a factor. The labor requirement summary must be performed operation by operation from the budgets. For example, the planner must decide whether he or she, as the operator, or a full-time hired laborer or a part-time laborer will disc each field, cultivate each

C-2   **BUDGET SHEET FOR CROP PRODUCTION**   For 19 ___   C-2

Crop to be Grown	Field					Seed			Fertilizer			Chemicals		
	No.	Acres	Yield Per Acre	Share %	Total Production	Variety	Rate per Acre	Total Quantity	Analysis	Rate per Acre	Total Quantity	Kind	Rate per Acre	Total Quantity

This form is copyrighted. It is a violation of the U.S. Copyright Law to reproduce it in any manner. To order forms write or call Century Communications Inc., 6201 Howard St., Niles, IL 60648; 708/647-1200.

Figure 7.5. Budget sheet for crop production, AFRA Schedule C-2.

Source: A.W. Oltmans, D.A. Klinefelter, and T.L. Frey, *Agricultural Financial Reporting and Analysis,* Century Communications, Inc, Niles, Ill., 1992.

crop, bale hay, and so on. The timing of each of these operations must also be listed, because there is little flexibility in when the operations can be performed, and labor cannot be transferred from month to month.

The operations summary can be translated to a complete whole-farm operating plan. The disaggregation of labor requirements by category can also be used as a plan of when labor of what type is needed. The operations summary must be interpreted carefully, however, before the labor hiring plan can be completed. The full-time labor needs must be rounded to 8- or 10-hour days, 5 or 6 days per week all year long, and priced accordingly. Direct wages, indirect wages, the employer's portion of the social security tax, and any other cash perquisites (such as health insurance) must be considered in the development of the labor budget. For part-time workers, similar adjustments must be made, to 8- or 10-hour days for day laborers, or to 8- or 10-hour days and 5- or 6-day weeks for summer employees. All cash commitments, especially for social security taxes, must be considered.

Fuel needs are a bit simpler to calculate than labor, because fuel can be purchased, stored, and used over several time periods. Total fuel needs by month can be estimated from the enterprise budgets if the operations have been allocated to a

Figure 7.6. Summary of crop costs, AFRA Schedule C-3.

Source: A.W. Oltmans, D.A. Klinefelter, and T.L. Frey, *Agricultural Financial Reporting and Analysis,* Century Communications, Inc, Niles, Ill., 1992.

particular time period. To translate these needs into a fuel purchasing schedule, recognize first that some fuel is likely to be included in the on-farm fuel tanks at the beginning of the year. Subtract January needs from January 1 inventory. If a deficit exists (A − B is less than zero in the worksheet shown in Figure 7.7), estimate purchases, probably a multiple of the storage tank capacity, and enter that amount as purchasing requirements. Complete the process by adding C + D = E to get the end-of-the-month inventory or the inventory for the beginning of the next month. The purchases column of the worksheet should then be transferred to Schedule C-3 and priced to get total fuel costs.

Other Purchased Supplies

Although not listed on Schedule C-3, other purchased inputs can be troublesome in developing a comprehensive cash flow plan. Feed purchasing schedules were discussed earlier in this chapter and repairs will be discussed in Chapter 8, but purchased inputs such as baling wire or twine should be listed on a worksheet for transfer to the cash flow budget. Budgeting for some supplies, however, can be overly tedious. Examples might include the need for postage stamps, business envelopes, pliers, grease, nails, and bolts. We know that these inputs will be needed, but budgeting for each conceivable item can quickly add up to demand more of the planner's time than can be justified. Rather than budgeting for each item, the planner

MONTH	INVENTORY, BEGINNING OF MONTH	MONTHLY FUEL NEEDS	DEFICIT OR SURPLUS	PURCHASES	INVENTORY END OF MONTH
	A	B	C	D	E
January					
February					
March					
April					
May					
June					
July					
August					
September					
October					
November					
December					

A - B = C
C + D = E

Figure 7.7. Gallons of machinery fuel budget.

would be better off to aggregate all of these typically small bills into a miscellaneous category. Make this category a "slush fund" for odd items and a margin for error in other input categories.

Past Records

The discussion in this section has taken the approach that we were starting the planning from scratch, almost as if we were planning an entirely new operation. While this approach is definitely useful to a new farm or ranch operator, it tends to ignore the experience accumulated by a veteran farmer or rancher. Past farm or ranch records can prove to be an invaluable substitute for some of the details of operations planning. However, the detailed planning process can surely give a great deal of insight to even a long-time veteran and can provide a good document for later control and analysis.

INVENTORY COSTS

Inventories of inputs can have significant cash flow impacts in several ways. First, inventories must be financed. Either cash resources that could have been used or invested elsewhere must be diverted, or credit resources that usually incur interest costs must be used, to finance inventory purchases. In the context of an overall cash flow plan, the various input marketing strategies developed through the processes described in Chapter 5 should be assessed once the overall plan is put together. Only then can we properly evaluate the cash use tradeoffs. For example, hay can usually be purchased throughout the year, but typically hay is produced and harvested in the summer and fed most heavily in the winter when grazing is least available. Consequently, hay follows a seasonal price pattern that displays season lowest prices in mid-summer and season highest prices in mid- to late winter.

Assume for the sake of convenience that hay prices reach a seasonal low of $100 per delivered ton on July 1 and a seasonal high of $125 on January 1. One marketing strategy might be to contract 100 percent of anticipated needs on July 1 to capture season-low prices. Another might be to make all purchases on January 1 to eliminate inventory carrying costs. A third alternative would be to make all purchases late in December rather than waiting until January. This third strategy adds an income tax management dimension to the problem. To evaluate the first two alternatives outside the overall cash flow plan (either to reduce the number of issues to juggle or just to have a starting point), compute total carrying costs for each alternative as follows: Price per ton + interest costs + insurance costs + storage costs + allowance for storage losses.

$$\text{Alternative 1:} \quad \$100 + 100 \times .12 \times \frac{6}{12} + 100 \times .01 + 100 \times .02 + 100 \times .04$$
$$= 100 + 3 + 1 + 2 + 4 = \$100$$

$$\text{Alternative 2:} \quad \$125$$

Insurance, shelter and storage costs must be added to interest costs to appropriately compare Alternative 1 to Alternative 2, because if hay is purchased in July, all storage losses and handling costs accrue to the buyer.

Another common inventory cost control problem relates to how much should be purchased at one time, especially for an input that is used rather steadily throughout the year. An economic order quantity (EOQ) level is found as:

$$EOQ = \sqrt{\frac{2FS}{CP}}$$

where EOQ = economic order quantity, the optimal amount to be ordered
   F = fixed costs of ordering, including in-transportation
   S = annual units needed
   C = annual carrying costs, expressed as a percentage of average inventory value
   P = purchase price per unit of inventory

EOQ might be applied to various farm and ranch commodities, including hay in a feedlot operation or dairy, purchased feed or medicines, or fuel for a year-round farming operation.

Consider an animal medicine example in which the medicine costs $20 per unit, has fixed costs of $25 per shipment/order, annual use of 1,500 units, and annual carrying costs equal to 10 percent of average inventory (to capture interest and losses). The EOQ would be approximately 200 units:

$$EOQ = \sqrt{\frac{2 \times 25 \times 1,500}{0.10 \times 20}} = \sqrt{\frac{75,000}{2}} = 194 \approx 200 \text{ units}$$

The EOQ computation works properly only if the input is used in an even pattern throughout the year and has a stable, predictable purchase price. These conditions obviously impose severe limitations on its applicability to agricultural production, in which many of the big ticket purchased inputs are not used or applied at a uniform rate throughout the year and have prices that fluctuate with market forces.

The limitations of models like the EOQ, which was developed for nonagricultural businesses, indicate even more strongly the need to develop an overall, consistent cash flow plan under which various input marketing strategies can be evaluated.

INVENTORY FLOWS AND PURCHASING SCHEDULES: SNAKE BELLY FARMS

Joe Farmer, in the spirit of a complete whole-farm analysis, decided to schedule all of his farm operations and input needs so that he could see all at once how well he has managed his operation in the past and how he might better control his costs in the future. Building from his enterprise budgets, Joe built his budget sheet for crop production and his summary of crop costs (Figures 7.8 and 7.9), his fuel, custom hire, and labor requirements summary (Figure 7.10), and fuel purchase schedule

(Figure 7.11).

    Joe built his inventory flow schedules first from his beginning balance sheet to derive beginning inventories and second from his production function analysis, whole-farm plan, and enterprise budgets to derive purchases, sales, death loss, feed use, and losses for all commodities. Figures 7.12 through 7.14 present Joe's inventory budgets for stocker steers, corn, and alfalfa (all other commodities produced were either not sold or were sold in the month harvested).

CROP DESC	PERCENT	FIELD #	ACRES	YIELD	TOTAL PROD
COTTONSEED	100	4	75.00	1200.00	90,000

CROP DESC	PERCENT	FIELD #	ACRES	YIELD	TOTAL PROD
SOYBEANS	100	2	0.00	45.00	0

	TYPE	DESCRIPTION	RATE/ACRE	UNIT	CONV FACT	TOTAL	UNIT
S = SEED	S	FORREST	20.00	LB	1.00	0.00	LBS
F = FERT	F	16-20-0-SOY	200.00	LB	2000.00	0.00	TONS
C = CHEM	C	DUAL 8EC-SOY	2.00	PT	8.00	0.00	GALS
	C	2,4-DB-SOY	3.37	PT10	80.00	0.00	GALS
	C	INSECT-SOY	1.00	APPL	1.00	0.00	APPL

CROP DESC	PERCENT	FIELD #	ACRES	YIELD	TOTAL PROD
CORN	100	3	0.00	170.00	0

	TYPE	DESCRIPTION	RATE/ACTE	UNIT	CONV FACT	TOTAL	UNIT
S = SEED	S	PIONEER	32000.00	K	80000.00	0.00	BAGS
F = FERT	F	UREA-CORN	435.00	LB	2000.00	0.00	TONS
C = CHEM	C	INSECT-CORN	2.00	APPL	1.00	0.00	APPL

CROP DESC	PERCENT	FIELD #	ACRES	YIELD	TOTAL PROD
COTTON	100	4	75.00	750.00	56,250

	TYPE	DESCRIPTION	RATE/ACRE	UNIT	CONV FACT	TOTAL	UNIT
S = SEED	S	DEKALB 565	20.00	LB	50.00	30.00	BAGS
F = FERT	F	16-20-0-COTT	200.00	LB	2000.00	7.50	TONS
C = CHEM	C	DUAL 8EC	1.00	APPL	1.00	75.00	ACRE
	C	ATRAZINE 4L	1.00	APPL	1.00	75.00	ACRE
	C	INSECT-COTT	1.00	APPL	1.00	150.00	ACRE

CROP DESC	PERCENT	FIELD #	ACRES	YIELD	TOTAL PROD
WHEAT	100	5	58.00	70.00	4,060

	TYPE	DESCRIPTION	RATE/ACRE	UNIT	CONV FACT	TOTAL	UNIT
S = SEED	S	COKER 916	100.00	LB	1.00	5800.00	LBS
F = FERT	F	18-46-0	108.70	LB	2000.00	3.15	TONS
C = CHEM	F	UREA 46-0-0	283.60	LB	2000.00	8.22	TONS
	C	INSECT-WHEAT	1.00	APPL	1.00	58.00	APPL

Figure 7.8. Budget sheet for corp production, Snake Belly Farms.

CROP DESC	PERCENT	FIELD #	ACRES	YIELD	TOTAL PROD
PASTURE	100	6	100.00	14.00	1,400

	TYPE	DESCRIPTION	RATE/ACRE	UNIT	CONV FACT	TOTAL	UNIT
S = SEED	F	16-20-0	100.00	LB	2000.00	5.00	TONS
F = FERT	S	GRASS SEED	30.00	LB	15.00	200.00	LBS
C = CHEM	F	16-20-0-PAST	100.00	LB	15.00	666.66	LBS

CROP DESC	PERCENT	FIELD #	ACRES	YIELD	TOTAL PROD
ALFALFA	100	7	300.00	7.00	2,100

	TYPE	DESCRIPTION	RATE/ACRE	UNIT	CONV FACT	TOTAL	UNIT
S = SEED	F	10-26-10	200.00	LBS	2000.00	30.00	TONS
F = FERT	C	INSECT-ALF	1.00	APPL	1.00	300.00	ACRE
C = CHEM	C	HERB-ALF	1.00	APPL	1.00	300.00	ACRE
	S	BALING TWINE	1190.00	FEET	4000.00	89.25	ROLL
	S	ALFALFA	30.00	LBS	6.00	1500.00	LBS
	C	INSECT - ALF E	1.00	APPL	6.00	50.00	ACRE

CROP DESC	PERCENT	FIELD #	ACRES	YIELD	TOTAL PROD
ALFALFA	100	8	20.00	5.00	100

	TYPE	DESCRIPTION	RATE/ACRE	UNIT	CONV FACT	TOTAL	UNIT
S = SEED	F	10-26-10	160.00	LBS	2000.00	1.60	TONS
F = FERT	C	INSECT-ALF	1.00	APPL	1.00	20.00	ACRE
C = CHEM	C	HERB-ALF	1.00	APPL	1.00	20.00	ACRE
	S	BALING TWINE	850.00	FEET	4000.00	4.25	ROLL
	S	ALFALFA	30.00	LBS	6.00	100.00	LBS
	C	INSECT - ALF E	1.00	APPL	6.00	3.33	ACRE

CROP DESC	PERCENT	FIELD #	ACRES	YIELD	TOTAL PROD
PASTURE	100	9	100.00	12.70	1,270

	TYPE	DESCRIPTION	RATE/ACRE	UNIT	CONV FACT	TOTAL	UNIT
S = SEED	S	GRASS SEED	30.00	LBS	15.00	200.00	LBS
F = FERT	F	16-20-0	100.00	LBS	2000.00	5.00	TONS
C = CHEM	F	16-20-0	100.00	LBS	15.00	666.66	LBS

CROP DESC	PERCENT	FIELD #	ACRES	YIELD	TOTAL PROD
PASTURE	100	5	80.00	3.80	304

Figure 7.8. (*continued*).

SUMMARY

The concept of inventory planning and control for farms and ranches is usually simpler than for manufacturing or retail firms and always adheres to a simple to understand equation: Sources = Uses. A typical sources and uses, or inventory flow, schedule takes the form of beginning inventory + purchases + production = on-farm uses (seed, feed, personal) + sales + losses + ending inventory. Most of the information needed for the inventory flow budgets has already been generated in the balance sheet, production function analysis, whole-farm plan, enterprise budgets, and marketing plans but still must be scheduled on budget sheets like AFRA cash flow schedules C-4 and C-5 to provide a physical flow check.

SNAKE BELLY

SEED	DESCRIPTION	PERCENT	QUANTITY	&/UNIT	TOTAL COST	MONTH
	DEKALB 565	100	30.0	20.00	600.00	4
	COKER 916	100	5800.00	.20	1160.00	8
	GRASS SEED	100	400.00	1.00	400.00	8
	BALING TWINE	100	93.50	26.00	2431.00	5
	ALFALFA	100	1600.00	2.50	4000.00	8
				TOTAL	8,591.00	

FERTILIZER	DESCRIPTION	PERCENT	QUANTITY	&/UNIT	TOTAL COST	MONTH
	UREA 46-0-0	100	8.22	230.00	1890.60	8
	18-46-0	100	3.15	320.00	1008.00	8
	16-20-0	100	10.00	180.00	1800.00	8
	16-20-0-PAST	100	1333.32	.09	119.99	8
	10-26-10	100	31.60	400.00	12640.00	8
	16-20-0-COTT	100	7.50	180.00	1350.00	4
				TOTAL	18,808.59	

CHEMICALS	DESCRIPTION	PERCENT	QUANTITY	&/UNIT	TOTAL COST	MONTH
	DUAL 8EC	100	75.00	9.00	675.00	5
	ATRAZINE 4L	100	75.00	10.00	750.00	7
	INSECT-COTT	100	150.00	12.50	1875.00	7
	INSECT-WHEAT	100	58.00	12.00	696.00	9
	INSECT-ALF	100	320.00	20.00	6400.00	4
	HERB-ALF	100	320.00	6.00	1920.00	6
	INSECT-ALF E	100	53.33	10.00	533.30	8
				TOTAL	12,849.30	

Figure 7.9. Summary of crop costs, Snake Belly Farms.

GAS, FUEL, OIL	DESCRIPTION	QUANTITY	$/UNIT	TOTAL COST	MONTH
	DIESEL	500.00	1.00	500.00	4
	DIESEL	500.00	1.00	500.00	5
	DIESEL	500.00	1.00	500.00	6
	DIESEL	1000.00	1.00	1000.00	7
	DIESEL	1500.00	1.00	1500.00	8
	DIESEL	500.00	1.00	500.00	9
	DIESEL	1500.00	1.00	1500.00	11
	NATURAL GAS	1568.00	4.65	7291.20	4
	NATURAL GAS	1638.00	4.65	7616.70	5
	NATURAL GAS	1743.00	4.65	8104.95	6
	NATURAL GAS	1841.00	4.65	8560.65	7
	NATURAL GAS	1985.00	4.65	9230.25	8
	NATURAL GAS	2109.00	4.65	9806.85	9
	NATURAL GAS	110.00	4.65	511.50	10
	OIL	25.00	4.00	100.00	4
	OIL	25.00	4.00	100.00	5
	OIL	25.00	4.00	100.00	6
	OIL	25.00	4.00	100.00	7
	OIL	50.00	4.00	200.00	8
	OIL	25.00	4.00	100.00	9
	OIL	50.00	4.00	200.00	11
			TOTAL	58,022.10	

CUSTOM MACHINE HIRE	DESCRIPTION	QUANTITY	$/UNIT	TOTAL COST	MONTH
	COMBINE WHEAT	58.00	26.40	1531.20	6
	HOE COTTON	75.00	20.00	1500.00	6
	GIN COTTON	75.00	84.75	6356.25	12
			TOTAL	9,387.45	

LABOR	DESCRIPTION	QUANTITY	$/UNIT	TOTAL COST	MONTH
	FULL TIME WORKER	1.00	850.00	850.00	1
	FULL TIME WORKER	1.00	850.00	850.00	2
	FULL TIME WORKER	1.00	850.00	850.00	3
	FULL TIME WORKER	1.00	850.00	850.00	4
	FULL TIME WORKER	1.00	850.00	850.00	5
	FULL TIME WORKER	1.00	850.00	850.00	6
	FICA DEPOSIT	1.00	900.00	900.00	6
	WITHHOLDING DEP	1.00	450.00	450.00	6
	FULL TIME WORKER	1.00	850.00	850.00	7
	FULL TIME WORKER	1.00	850.00	850.00	8
	FULL TIME WORKER	1.00	850.00	850.00	9
	FULL TIME WORKER	1.00	850.00	850.00	10
	FULL TIME WORKER	1.00	850.00	850.00	11
	FULL TIME WORKER	1.00	850.00	850.00	12
	FICA DEPOSIT	1.00	900.00	900.00	12
	WITHHOLDING DEP	1.00	450.00	450.00	12
			TOTAL	12,900.00	

Figure 7.10. Fuel, labor requirements, and custom hire summary, Snake Belly Farms.

MONTH	INVENTORY BEGINNING OF MONTH A	MONTHLY FUEL NEEDS B	DEFICIT OR SURPLUS C	PURCHASES D	INVENTORY, END OF MONTH E
January	500	100	400	0	400
February	400	100	300	0	300
March	300	100	200	0	200
April	200	386	(186)	500	314
May	314	567	(253)	500	247
June	247	658	(411)	500	89
July	89	658	(569)	1,000	431
August	431	1,426	(995)	1,500	505
September	505	707	(202)	500	298
October	298	100	(198)	0	198
November	198	1,268	(1,070)	1,500	430
December	430	100	330	0	330

Figure 7.11. Fuel needs summary, Snake Belly Farms.

Month	Beginning Inventory	Purchases	Died	Sales	Ending Inventory
1	540				540
2	540				540
3	540				540
4	540				540
5	540				540
6	540			540	0
7	0				0
8	0				0
9	0				0
10	0	550			550
11	550				550
12	550		10		540
Total	540	550	10	540	540

Figure 7.12. Stocker steer inventory schedule.

Month	Beginning Inventory	Purchases	Feed Use	Sales	Ending Inventory
1	2,000		335		1,665
2	1,665		335		1,330
3	1,300		335	326	669
4	669		335		334
5	334		334		0
6	0				0
7	0				0
8	0				0
9	0				0
10	0	1,026	342		684
11	684		342		342
12	342		342		0
Total	2,000	1,026	10	326	0

Figure 7.13. Corn inventory schedule.

Month	Beginning Inventory	Produced	Feed Use	Sales	Ending Inventory
1	100		8	81	11
2	11		7		4
3	4		4		0
4	0				0
5	0	200			200
6	200	600			800
7	800	500			1,300
8	1,300	500			1,800
9	1,800	400			2,200
10	2,200			1,022	1,178
11	1,178				1,178
12	1,178			1,078	100
Total	100	2,200	19	2,181	100

Figure 7.14. Alfalfa inventory schedule.

NOTES

1. See, for example, the economic order quantity and inventory control discussion in E.F. Brigham and L.C. Gapenski, *Financial Management: Theory and Practice*, 6th ed., The Dryden Press, Chicago, Ill., 1991, Chapter 21; or W.J. Morse, J.R. Davis, and A.L. Hartgraves, *Management Accounting*, 3rd ed., Addison-Wesley Publishing Company, Reading, Mass., 1991, Chapter 14.

2. We may need to project losses due to short pieces left at the end of a roll, the inevitable use of baling wire to repair machinery, or possible losses due to theft.

RECOMMENDED READINGS

Robert B. Schwart, J. Milt Holcomb, and Allan G. Mueller, *Mechanics of Farm Financial Planning*, Circular 1042, Cooperative Extension Service, University of Illinois, Urbana, Ill., October 1971.

Timothy G. Baker and John R. Brake, *Cash Flow Analysis of the Farm Business*, Extension Bulletin E-911, Cooperative Extension Service, Michigan State University, East Lansing, Mich., October 1975.

A. Gene Nelson and Thomas L. Frey, *You and Your Cash Flow*, Century Communications, Inc., Skokie, Ill., 1988.

Eugene F. Brigham and Louis C. Gapenski, *Financial Management: Theory and Practice,* 6th ed., The Dryden Press, Chicago, Ill., 1991, Chapter 19.

Wayne J. Morse, James R. Davis, and Al L. Hartgraves, *Management Accounting,* 3rd ed., Addison-Wesley Publishing Company, Reading, Mass., 1991, Chapter 14.

CHAPTER 8

Investment, Repair, Debt Service, and Other Expenditure Planning

The last chapter in this section concerning the development of component plans brings together several seemingly unrelated planning tasks. We do this not to suggest that they are unimportant but rather to recognize that these plans either take very little effort or depend in a very major way on what happens throughout the year. We begin with debt service plans for existing amortized loans and proceed to repairs and capital expenditures, family living needs, and tax liabilities.

DEBT SERVICE PLANS

Planning for intermediate and long-term loan payments must be broken into two parts: the loans that exist at the beginning of the year and the new loans that must be taken out during the year to finance new purchases or to refinance existing debt. At this point in our discussion we will ignore new intermediate and long-term loans (whether they relate to new purchases or to refinancing) and delay that discussion until Chapter 9. Developing the component plans for the servicing of existing debt is, for a change, easy and straightforward, if the planner has the following two things in hand:

1. A completed beginning-of-the-year balance sheet
2. An understanding of loan repayment and interest calculation methods

All of the information needed to project and plan for debt service needs can be extracted from Schedule 12 of the AFRA balance sheet (refer to Figure 2.11 for a display of the schedule). That schedule lists all existing scheduled notes and the repayment terms for those notes. A current principal balance, appropriate periodic interest rate, and payment date are the necessary and sufficient pieces of information to calculate the interest portion of the loan payment due within the year and the required timing of the payment. The principal due within 12 months is the necessary information to project the principal portion of the loan payment; it is sufficient also if the loan is amortized on annual installments. More work will be required to determine the timing of the principal portions if more than one payment is scheduled during the year.[1] Most of this information can be requested from the lender—ask for an amortization schedule for the life of the loan.

Using a separate line of AFRA Cash Flow Schedule C-7 (Figure 8.1) for each existing intermediate and long-term loan, the timing of loan repayment requirements can be determined easily, because they are set by the terms of the loan. For new loans, these terms are negotiable. For cash flow purposes, the total payment is sufficient, but for later projection of the income statement and balance sheet, it is necessary to know both the interest and principal portions.

C-7 TERM DEBT LOAN PAYMENTS For 19___ C-7

Beginning Balance on Previous Loans	Due Date	Interest Rate	Payments	Jan.	Feb.	March	April	May	June	July	August	Sept.	Oct.	Nov.	Dec.	Ending Principal Balance
$			Principal													$
			Interest													
NEW LOANS																

This form is copyrighted. It is a violation of the U.S. Copyright Law to reproduce it in any manner. To order forms write or call Century Communications Inc., 6201 Howard St., Niles, IL 60714; 708/647-1200 or Doane Agricultural Services Co., 11701 Borman Dr., St. Louis, MO 63146; 314/569-2700.

Figure 8.1. Term debt loan payments, AFRA Schedule C-7.

Source: A.W. Oltmans, D.A. Klinefelter, and T.L. Frey, *Agricultural Financial Reporting and Analysis,* Century Communications, Inc, Niles, Ill., 1992.

REPAIRS

Projecting repair expenditures could be an easy or an impossible task. We know that some repairs of machinery and fixed improvements will be needed, but how much and when? Good planning for repairs will draw on engineering estimates (usually summarized on enterprise budgets) and past farm or ranch records for a guideline of how many and when repairs can be expected. But those historical records will show extreme variability from year to year, as well as dependence on age of equipment, crop acres, and pure chance. We can probably predict timing better than we can the amount: simply ask when the machine is most needed and assume that is when the major repairs will be needed. Good repair planners are believers in Murphy's First and Fourth laws: "If anything can go wrong, it will," and "If there is a possibility of several things going wrong, the one that will go wrong first will be the one that will do the most damage." On a more serious level, some major repairs can be estimated based on current knowledge of the condition of each machine or improvement. For example, the repair or overhaul of a weak irrigation pump or motor can be projected based on last year's performance, or the operator may know that all of the worn shovels on a row cultivator must be replaced before another season rolls around. However, most repairs are very unpredictable.

To estimate repair costs, the best approach may well be to estimate the total annual repair cost based on past records or enterprise budgets and then to spread that total over the months in the primary growing season, thus creating a contingency or slush fund. Record these estimates on AFRA Cash Flow Schedule C-6 (Figure 8.2), and label scheduled major repairs and the contingency fund separately.

CAPITAL EXPENDITURES

Planning for capital expenditures is related to and much like planning for repairs. A particular machine might break down seriously enough to warrant replacement rather than further repair. Unless we have specific knowledge of the age and condition of each machine and improvement, it is almost impossible to project these instances. Few farmers and ranchers replace capital assets on a regular rotation, i.e., replace the primary tractor every six years. Rather, replacement is usually based on immediate need caused by an unrepairable breakdown or on the availability of extra cash generated by favorable output prices. Some capital expenditures are, however, predictable. Expansions of certain enterprises may require additional capital outlays, or entering new enterprises may require the purchase of a specialized piece of equipment, such as a precision planter for vegetables. Record the total dollar value (not the down payment if it is to be financed) of planned capital expenditures on Schedule C-6 (Figure 8.2) based on the best guess of when that purchase will take place. We will discuss the financing of these purchases and expenditures in Chapter 9.

No capital expenditure, regardless of type or even amount, should be made without a thorough analysis. A long-term cash flow forecast, analyzed in a capital budgeting framework, is essential to good decision making. That capital budgeting analysis must be consistent with the long-term goals of the business and its owners, as well as consistent with the long-term financial strength of the business. Capital budgeting will be discussed further in Chapter 12.

FAMILY LIVING EXPENDITURES

Planning for family living expenditures is a controversial subject because it involves the family's goals, wants, and desires. Furthermore, it is complicated because it draws on a separate set of records and another checking account. Nonetheless, family living budgeting and planning are serious and necessary projects, but they may be best approached as a totally separate planning activity from the business planning process. Not only should the family living needs be considered, but off-farm income, investment income, investment expenditures, and retirement plans should also be included.

The AFRA Cash Flow Statement includes a family living budget (Schedule C-8, Figure 8.3) that can be completed in whatever detail is desired. Based on past records or on nonfarm planning activities, it may be sufficient to complete only the bottom line, the gross family living withdrawals. If this is done, there should be greater documentation in other sources.[2]

C-6	CAPITAL EXPENDITURES AND REPAIRS			For 19_____			C-6
				Purchases & Capital Repairs		Operating Repairs	
Description	Number or Quantity	Month	Machinery and Equipment	Buildings and Improvements	Machinery and Equipment	Buildings and Improvements	
			$	$	$	$	

This form is copyrighted. It is a violation of the U.S. Copyright Law to reproduce it in any manner. To order forms write or call Century Communications Inc., 6201 Howard St., Niles, IL 60714; 708/647-1200 or Doane Agricultural Services Co., 11701 Borman Dr., St. Louis, MO 63146; 314/569-2700.

Figure 8.2. Capital expenditures and repairs, AFRA Schedule C-6.
Source: A.W. Oltmans, D.A. Klinefelter, and T.L. Frey, *Agricultural Financial Reporting and Analysis,* Century Communications, Inc, Niles, Ill., 1992.

TAX PAYMENTS

One more important but nebulous planning activity remains: projecting income and social security tax payments attributable to the business owner-operator and family unit. To plan cash needs for the year without considering the tax liability due in the first few months on last year's income or due with quarterly estimates would be foolish. Last year's income statement and cash flow summary, along with a tax estimate worksheet and appropriate tax rate schedules, should be enough information to project the tax due on last year's activities. That number should also be on the beginning of the year balance sheet. Use a tax estimate like the one shown in Figure 8.4 and tax rate schedules from Form 1040-ES (Figure 8.5) to estimate this tax liability, or request a preliminary estimate from the farm or ranch tax preparer. If the business satisfies the Internal Revenue Service definition of a farm (two-thirds or more of gross income is attributable to farming activities), then quarterly estimates are not required; however, if the business does not satisfy this definition, attention must be paid to the instructions on the current year's Form 1040-ES.[3,4]

C-8				FAMILY LIVING BUDGET							For 19____		C-8
	Jan	Feb	March	April	May	June	July	August	Sept.	Oct.	Nov.	Dec.	TOTAL
Food													
Household operating expenses													
Household equipment and furnishings													
House repairs													
Rent													
Clothing													
Personal													
Entertainment and recreation													
Education and reading materials													
Medical care and drugs													
Church and charity													
Personal gifts													
Utilities (non-farm)													
Transportation and auto (non-farm)													
Personal and recreational vehicle purchases													
Medical and disability insurance													
Life insurance													
Total Family Living Withdrawals													

This form is copyrighted. It is a violation of the U.S. Copyright Law to reproduce it in any manner. To order forms write or call Century Communications Inc., 6201 Howard St., Niles, IL 60714; 708/647-1200 or Doane Agricultural Services Co., 11701 Borman Dr., St. Louis, MO 63146; 314/569-2700.

Figure 8.3. Family living budget, AFRA Schedule C-8.
Source: A.W. Oltmans, D.A. Klinefelter, and T.L. Frey, *Agricultural Financial Reporting and Analysis,* Century Communications, Inc, Niles, Ill., 1992.

```
                                                            Estimated   Estimated
                                               Amount to    Rest of     Year's
                                               Date         Year        Total
RECEIPTS                                       ---------------------------------------
Sales of products raised(*) and miscellanous receipts:
    Cattle, hogs, sheep and wool . . . . . . . . . . . . .  $ _____   _____   _____
    Poultry, eggs and dairy products . . . . . . . . . . .    _____   _____   _____
    All crop sales . . . . . . . . . . . . . . . . . . . .    _____   _____   _____
    Custom work, prorations, refunds and
       agriculture program payments . . . . . . . . . . . .    _____   _____   _____
Total sales and other income . . . . . . . . . . . . . (1)    _____   _____   _____
Sales of purchased market livestock. . . .(**)$ _____
Cost of purchased market livestock . . . .(***)$ _____
Gross profits on sales of purchased livestock (sales - cost) . . . (2)
Gross farm profits (Item 1 + 2). . . . . . . . . . . . . . . (3)                   _____


EXPENSES
    Labor hired . . . . . . . . $ _____    Veterinary, medicine . $ _____
    Repairs, maintenance. . . . .   _____   Gasoline, fuel, oil. .    _____
    Interest. . . . . . . . . .     _____   Storage, warehousing .    _____
    Rent. . . . . . . . . . . .     _____   Taxes. . . . . . . . .    _____
    Feed purchased. . . . . . .     _____   Insurance. . . . . . .    _____
    Seed, plants purchased. . .     _____   Utilities. . . . . . .    _____
    Fertilizers . . . . . . . .     _____   Freight, trucking. . .    _____
    Chemicals . . . . . . . . .     _____   Conservation expenses.    _____
    Machine hire. . . . . . . .     _____   Other. . . . . . . . .    _____
    Supplies purchased  . . . .     _____   Other. . . . . . . . .    _____
    Breeding fees. . . . . . . .    _____   Other. . . . . . . . .    _____


Total cash expenses. . . . . . . . . . . . . . . . . . . (4) $ _____   _____   _____
Depreciation on machinery, improvements,
    dairy and breeding stock . . . . . . . . . . . . . . . (5)                      _____
Total deductions (Item 4 + 5). . . . . . . . . . . . . . (6)   _____
Self employment income
    (Item 3 - 6) . . . . . . . . . . . . . . . . . . . (**)(7)                      _____
Net taxable gain from Schedule D (sales of dairy
    and breeding stock, machinery and other
    capital exchanges) . . . . . . . . . . . . . . . . . (8)                        _____
Taxable non-business income. . . . . . . . . . . . . . . (9)   _____   _____   _____
Adjusted gross income (Item 7 + 8 + 9) . . . . . . . . . (10)  _____   _____   _____
Less:    Excess itemized deductions,
    from Schedule A . . . . . . . . . . . . . . $ _____
Personal exemptions. . . . . . . . . . . . . . $ _____
Total non-business deductions and exemptions . . . . . . . (11) _____  _____   _____
Taxable income (Item 10 - 11). . . . . . . . . . . . . . (12)  _____   _____   _____
Estimated income tax (from applicable tax
    rate schedules). . . . . . . . . . . . . . . . . . . (13)                       _____
Estimated self employment tax
    (Item 7 x current rate). . . . . . . . . . . . . . . (14)                       _____


TOTAL TAX (Item 13 + 14) . . . . . . . . . . . . . . . . (15)                       _____

Credits:  allowable investment credit and
    carryover, gas tax, income tax withheld
    and estimated tax paid . . . . . . . . . . . . . . . (16)                       _____

Estimated tax due (Item 15 - 16) . . . . . . . . . . . . (17)                       _____
```

----
* Omit for accrual method.
** For accrual method, adjust for change in inventory and new livestock purchases.
*** For accrual method, includes sales of all livestock.

Figure 8.4. Income and social security tax worksheet.

1993 Estimated Tax Worksheet (keep for your records)

1	Enter amount of adjusted gross income you expect in 1993	**1**	
2	• If you plan to itemize deductions, enter the estimated total of your itemized deductions. **Caution:** If line 1 above is over $108,450 ($54,225 if married filing separately), your deduction may be reduced. See Pub. 505 for details. • If you do not plan to itemize deductions, see **Standard Deduction for 1993** on page 2, and enter your standard deduction here.	**2**	
3	Subtract line 2 from line 1	**3**	
4	Exemptions. Multiply $2,350 by the number of personal exemptions. If you can be claimed as a dependent on another person's 1993 return, your personal exemption is not allowed. **Caution:** If line 1 above is over $162,700 ($135,600 if head of household; $108,450 if single; $81,350 if married filing separately), get Pub. 505 to figure the amount to enter	**4**	
5	Subtract line 4 from line 3	**5**	
6	**Tax.** Figure your tax on the amount on line 5 by using the 1993 Tax Rate Schedules on page 2. DO NOT use the Tax Table or the Tax Rate Schedules in the 1992 Form 1040 or Form 1040A instructions. **Caution:** If you have a net capital gain and line 5 is over $89,150 ($76,400 if head of household; $53,500 if single; $44,575 if married filing separately), get Pub. 505 to figure the tax	**6**	
7	Additional taxes (see line 7 instructions)	**7**	
8	Add lines 6 and 7	**8**	
9	Credits (see line 9 instructions). Do not include any income tax withholding on this line . . .	**9**	
10	Subtract line 9 from line 8. Enter the result, but not less than zero	**10**	
11	Self-employment tax. Estimate of 1993 net earnings from self-employment $...................; if **$57,600 or less,** multiply the amount by .153; if **more than $57,600,** see line 11 instructions for the amount to enter. **Caution:** If you also have wages subject to social security or Medicare tax, get Pub. 505 to figure the amount to enter	**11**	
12	Other taxes (see line 12 instructions)	**12**	
13a	Add lines 10 through 12	**13a**	
b	Earned income credit and credit from **Form 4136**	**13b**	
c	Subtract line 13b from line 13a. Enter the result, but not less than zero. **THIS IS YOUR TOTAL 1993 ESTIMATED TAX** . . . . ▶	**13c**	
14a	Multiply line 13c by 90% (66⅔% for farmers and fishermen) . . . **14a** _____		
b	Enter 100% of the tax shown on your 1992 tax return . . . . **14b** _____		
	Caution: If 14b is **smaller** than 14a **and** line 1 above is over $75,000 ($37,500 if married filing separately), stop here and see **Limit on Use of Prior Year's Tax** on page 1 before continuing.		
c	Enter the **smaller** of line 14a or 14b. **THIS IS YOUR REQUIRED ANNUAL PAYMENT TO AVOID A PENALTY** . . . . ▶	**14c**	
	Caution: Generally, if you do not prepay at least the amount on line 14c, you may owe a penalty for not paying enough estimated tax. To avoid a penalty, make sure your estimate on line 13c is as accurate as possible. Even if you pay the required annual payment, you may still owe tax when you file your return. If you prefer, you may pay the amount shown on line 13c. For more details, get Pub. 505.		
15	Income tax withheld and estimated to be withheld during 1993 (including income tax withholding on pensions, annuities, certain deferred income, etc.)	**15**	
16	Subtract line 15 from line 14c. (**Note:** If zero or less, or line 13c minus line 15 is less than $500, stop here. You are not required to make estimated tax payments.)	**16**	
17	If the first payment you are required to make is due April 15, 1993, enter ¼ of line 16 (minus any 1992 overpayment that you are applying to this installment) here and on your payment voucher(s)	**17**	

Figure 8.5. IRS form 1040-ES.

1993 Tax Rate Schedules

Caution: *Do not use these Tax Rate Schedules to figure your 1992 taxes. Use only to figure your 1993 estimated taxes.*

Single—Schedule X

If line 5 is: Over—	But not over—	The tax is:	of the amount over—
$0	$22,100	.........15%	$0
22,100	50,500	$3,315.00 + 28%	22,100
53,500	....	12,107.00 + 31%	53,500

Head of household—Schedule Z

If line 5 is: Over—	But not over—	The tax is:	of the amount over—
$0	$29,600	.........15%	$0
29,600	76,400	$4,440.00 + 28%	29,600
76,400	....	17,544.00 + 31%	76,400

Married filing jointly or Qualifying widow(er)—Schedule Y-1

If line 5 is: Over—	But not over—	The tax is:	of the amount over—
$0	$36,900	.........15%	$0
36,900	89,150	$5,535.00 + 28%	36,900
89,150	....	20,165.00 + 31%	89,150

Married filing separately—Schedule Y-2

If line 5 is: Over—	But not over—	The tax is:	of the amount over—
$0	$18,450	.........15%	$0
18,450	44,575	$2,767.50 + 28%	18,450
44,575	....	10,082.50 + 31%	44,575

Figure 8.5. *(continued).*

OTHER EXPENDITURES: SNAKE BELLY FARMS

Picking up from his beginning balance sheet (Figure 3.2), Joe Farmer was able to easily plan his existing debt service requirements as shown in Figure 8.6. He based his repair and capital expenditure plans (Figure 8.7) on last year's records, the enterprise budgets developed and discussed in Chapter 6 (see specifically Table 6.6), and personal experience with each machine. Specifically, he plans on overhauling his 96-horsepower tractor in February before the heavy field work requirements begin, he estimates normal repairs to be $9,129, and he plans to replace his pickup in the fall. Furthermore, Joe plans on demolishing dilapidated grain bins and reworking fence lines to fit the stocker operation. Building maintenance is estimated to be $900. Joe and Mary have always attempted to control their family budget as tightly as the business budget and usually attempt to plan their family living needs before beginning the business planning process. The results of their family budget planning are shown in Figure 8.8. Finally, Joe used last year's records to develop an estimate of 19X0 income and social security taxes due in February 19X1 (Figure 8.9).

SUMMARY

Planning for this last group of cash flow can be surprisingly easy or virtually impossible to do reasonably. Much of the information comes from previously developed documents or outside sources (e.g., debt service, family living, and tax estimates). The rest of the information (repairs and capital expenditures) must come from past records, enterprise budgets, and a strong dose of experience and judgment. Furthermore, some of the capital expenditure plans must come from the long-run goals for expansion or movement into new enterprises.

BEGINNING BALANCE	DATE	INT RATE	PMNT	JAN	FEB	MAR	APR	MAY	JUN	JUL	AUG	SEPT	OCT	NOV	DEC	ENDING BALANCE
165,000	06/01/X2	12.00	PRIN	0	0	0	0	0	165,000	0	0	0	0	0	0	0
			INT	0	0	0	0	0	13,200	0	0	0	0	0	0	
43,956	12/01/X2	14.00	PRIN	0	0	0	0	0	0	0	0	0	0	0	12,780	31,176
			INT	0	0	0	0	0	0	0	0	0	0	0	6,154	
7,114	12/15/X2	13.00	PRIN	0	0	0	0	0	0	0	0	0	0	0	3,340	3,774
			INT	0	0	0	0	0	0	0	0	0	0	0	925	
935,925	6/01/X2	9.00	PRIN	0	0	0	0	0	0	0	0	0	0	0	25,313	910,612
			INT	0	0	0	0	0	0	0	0	0	0	0	84,233	
151,995		TOTAL	PRIN	0	0	0	0	0	165,000	0	0	0	0	0	41,433	945,562
		TOTAL	INT	0	0	0	0	0	13,200	0	0	0	0	0	91,312	

Figure 8.6. Term debt loan payment schedule, Snake Belly Farms.

DESCRIPTION	QUANTITY	MONTH FROM	MONTH TO	PURCHASES AND CAPITAL REPAIRS		OPERATING REPAIRS	
				MACHINE/EQUIP	BUILD/IMPROVE	MACHINE/EQUIP	BUILD/IMPROVE
Major Improvement Repairs							
Demolish grain bins	1	6	6	0	0	0	1,000
Remove portion of fence	63	7	7	0	0	0	378
Repair internal fence	100	7	7	0	0	0	300
Capital Improvements							
Internal fencing	371	7	7	0	3,710	0	0
Bldg Maintenance & Repair	1	1	12	0	0	0	900
New Pickup	1	10		15,000	0	0	0
Tractor Overhaul	1	3	3	2,000	0	0	0
Equipment Repair Estimate	1	1	12	0	0	9,129	0
TOTALS				17,000	3,710	9,129	2,578

Figure 8.7. Capital expenditures and repairs, Snake Belly Farms.

DESCRIPTION	JAN	FEB	MAR	APRIL	MAY	JUNE	JULY	AUG	SEPT	OCT	NOV	DEC	TOTAL
FOOD	400	400	400	400	400	400	400	400	400	400	400	400	4,800
HOUSEHOLD OPER. EXP	300	300	300	300	300	300	300	300	300	300	300	300	3,600
HOUSEHOLD EQUIP/FURN	0	0	0	0	0	0	0	0	0	0	0	0	0
HOUSE REPAIRS	0	0	0	0	0	0	0	0	0	0	0	0	0
RENT	0	0	0	0	0	0	0	0	0	0	0	0	0
CLOTHING	100	100	100	100	100	100	100	100	100	100	100	100	1,200
PERSONAL	150	150	150	150	150	150	150	150	150	150	150	150	1,800
ENTERTAIN/RECREATION	25	25	25	25	25	25	25	25	25	25	25	25	300
EDUC/READ MATERIALS	0	0	0	0	0	0	0	0	0	0	0	0	0
MEDICAL CARE & DRUGS	25	25	25	25	25	25	25	25	25	25	25	25	300
CHURCH AND CHARITY	100	100	100	100	100	100	100	100	100	100	100	100	1,200
PERSONAL GIFTS	0	0	0	0	0	0	0	0	0	0	0	0	0
UTILITIES (NON-FARM)	75	75	75	75	75	75	75	75	75	75	75	75	900
TRAN/AUTO (NON-FARM)	75	75	75	75	75	75	75	75	75	75	75	75	900
PRSNL/REC VHCL PURCH	0	0	0	0	0	0	0	0	0	0	0	0	0
MEDICAL/DISABIL INS	150	150	150	150	150	150	150	150	150	150	150	150	1,800
LIFE INSURANCE	100	100	100	100	100	100	100	100	100	100	100	100	1,200
ADD. TO PRSNL INVEST	0	0	0	0	0	0	0	0	0	0	0	0	0
ADD. TO RETIRE ACCT	0	0	0	0	0	0	0	0	0	0	0	0	0
FAMILY LIVING	1,500	1,500	1,500	1,500	1,500	1,500	1,500	1,500	1,500	1,500	1,500	1,500	18,000

Figure 8.8. Family living budget, Snake Belly Farms.

	Amount to Date	Estimated Rest of Year	Estimated Year's Total
RECEIPTS			
Sales of products raised(*) and miscellaneous receipts:			
Cattle, hogs, sheep and wool . . . . $	0	0	0
Poultry, eggs and dairy products	0	0	0
All crop sales	124,967	92,000	216,967
Custom work, prorations, refunds and agricultural program payments	12,187	10,000	22,187
Total sales and other income . . . . (1)	127,154	102,000	229,154
Sales of purchased market livestock. . . . (**)$ 137,412			
Cost of purchased market livestock . . . .(***)$ 166,400			
Gross profits on sales of purchased livestock (sales - cost) . . . (2)			71,012
Gross farm profits (Item 1 + 2). . . . . (3)			310,166

EXPENSES

Labor hired . . . . $	10,000	Veterinary, medicine . $	3,900
Repairs, maintenance.	6,210	Gasoline, fuel, oil. .	57,130
Interest.	14,196	Storage, warehousing .	0
Rent.	2,152	Taxes.	2,046
Feed purchased.	3,967	Insurance.	3,150
Seed, plants purchased. . . .	8,235	Utilities.	1,810
Fertilizers	19,120	Freight, trucking. . .	0
Chemicals	14,167	Conservation expenses.	0
Machine hire.	3,247	Other.	0
Supplies purchased	4,120	Other.	0
Breeding fees.	0	Other.	0

	Amount to Date	Estimated Rest of Year	Estimated Year's Total
Total cash expenses. . . . . (4) $	153,450	110,000	263,450
Depreciation on machinery, improvements, dairy and breeding stock . . . . (5)			52,040
Total deductions (Item 4 + 5). . . . . (6)			315,490
Self employment income (Item 3 - 6) . . . . (**)(7)			-5,324
Net taxable gain from Schedule D (sales of dairy and breeding stock, machinery and other capital exchanges) . . . . (8)			0
Taxable non-business income. . . . . (9)	11,667	2,333	14,000
Adjusted gross income (Item 7 + 8 + 9) . . . .(10)			8,676
Less: Excess itemized deductions, from Schedule A . . . . $ 5,000			
Personal exemptions. . . . . $ 4,320			
Total non-business deductions and exemptions . . . . . .(11)			9,320
Taxable income (Item 10 - 11). . . . .(12)			-644
Estimated income tax (from applicable tax rate schedules). . . . .(13)			0
Estimated self employment tax (Item 7 x current rate). . . . .(14)			0
TOTAL TAX (Item 13 + 14) . . . .(15)			0
Credits: allowable investment credit and carryover, gas tax, income tax withheld and estimated tax paid . . . .(16)			0
Estimated tax due (Item 15 - 16) . . . .(17)			0

----

* Omit for accrual method.

** For accrual method, adjust for change in inventory and new livestock purchases.

*** For accrual method, includes sales of all livestock.

Figure 8.9. Income and social security tax estimate worksheet, Snake Belly Farms.

NOTES

1. Refer to an agricultural finance text or J.D. Libbin and L.B. Catlett, *Farm and Ranch Financial Records*, Macmillan, New York, N.Y., 1987, Chapter 8, for detailed analysis of loan repayment and interest calculation methods.

2. A complete discussion of family finance and budgeting is beyond the scope of this book. Although family involvement in farm or ranch business planning would be extremely valuable, that involvement is absolutely necessary for family living expenditure planning. Interested readers should consult a personal finance book like J.B. Penson, D.R. Levi, and C.J. Nixon, see recommended readings, and maintain a family record book.

3. For a more complete IRS definition of a farm, see Chapter 1 of the current *Farmer's Tax Guide*, IRS Publication 225.

4. Like family living budgeting, income tax planning is beyond the scope of this book. For an introduction to farm and ranch income tax reporting see J.D. Libbin and L.B. Catlett, *Farm and Ranch Financial Record*, Macmillan, New York, N.Y., 1987.

RECOMMENDED READINGS

John B. Penson, Jr., Donald R. Levi, and Clair J. Nixon, *Personal Finance,* Prentice-Hall, Englewood Cliffs, N.J., 1982.

IRS, *Farmers Tax Guide,* Publication 225, Internal Revenue Service, Washington, D.C.

James D. Libbin, *Estimating Farm and Ranch Income Tax,* Guide Z-805, Cooperative Extension Service, New Mexico State University, Las Cruces, September 1984.

David Niederkorn, "'Incidentals' Can Be a Major Incident for Cash Flow Management," *Money Sense,* John Deere, Moline, Ill., Spring 1987.

James D. Libbin and Lowell B. Catlett, *Farm and Ranch Financial Records,* Macmillan, New York, N.Y., 1987.

PART III

Compilation and Use of a Total Plan

Once all of the component plans, or at least their first drafts, are completed (with the recognized exception of financing plans), we are in a position to put all of the plans together into a complete cash flow budget. This point is where the "fun" really begins as we must use all of the available tools to finance the operation by borrowing and repaying short-term loans; borrowing to finance capital expenditures over a period of years; and iteratively adjusting marketing, production, inventory, other plans; and maybe even our short-term goals to prepare the most realistic and "best" cash flow plan for the year.

A single cash flow plan based on expected yields, prices, and costs simply will not recognize the realities of changing conditions that must be faced by every farmer and rancher. Chapter 10 will direct attention to ways of recognizing the impact of risk in the only framework that really matters: its effect on cash flow.

Part III will close with Chapter 11 and its recognition that a cash flow plan developed for credit application and discarded, rather than being used

throughout the year, is a sterile document and a nearly totally wasted effort. One of the major roles of effective business management is control: control of business activities, control of profit, and most immediately, control of cash.

CHAPTER 9

Putting the
Plan Together

The end result of the process of cash flow planning is to have a complete, comprehensive cash flow plan. To remind ourselves of where we are heading, we repeat Figure 1.3 in Figure 9.1. In Chapters 4 through 8, we discussed almost all of the component plans necessary to complete our comprehensive plan. Chapter 4 helped us determine where we want to go; although there is no identifiable section in Figure 9.1 labeled Goal Plans, it should be clear that our goals have influenced our choices all the way through the remainder of the planning process and should continue to influence our choices as we finish the process. The marketing and production planning topics of Chapters 5 and 6 clearly determine the receipts and expenditures sections in

	Period 1	Period 2	...
Beginning Cash Balance		A	
+ Receipts			
= Total Cash Available			
− Expenditures			
= Cash Available Less Cash Required			
+ Inflows from Savings			
+ Cash Borrowed			
− Short-term Loan Repayments			
− Outflows to Savings Accounts			
= Ending Cash Balance	A		

Figure 9.1. General structure of a cash flow budget.

Figure 9.1, along with the inventory plans of Chapter 7 and the remaining plans of Chapter 8.

Consequently, the first step in putting together a comprehensive, complete cash flow budget is to assemble all of the components described in Chapters 4 through 8 and formalized into schedules in Chapters 6, 7, and 8. The next step is to adjust cash outflows to match cash inflows or to find ways to finance the operation (the four lines between Cash Available Less Cash Required and Ending Cash Balance in Figure 9.1). Those four lines generally show our opportunities to finance the operation during periods of cash deficits: self-financing through the transfer of funds from a savings account to the business checking account or the borrowing of short-term funds. If we borrow short-term or operating funds, we have generated an obligation to repay those funds, usually within the calendar (or operations) year, or, if we self-finance, we must replenish the savings account for use in the next year.

AFRA CASH FLOW BUDGET

To begin the process of putting the cash flow plan together, let's begin with a quick review of the AFRA cash flow budget, shown in Figure 9.2. The AFRA cash flow budget has the same basic structure as the simplified example of Figure 9.1. It begins with the Beginning Cash Balance (Line 1), and adds Operating Receipts or sales of crops and livestock raised primarily for sale and other normal or recurring farm and ranch income (Lines 2 through 8), Capital Receipts or sales of capital assets (Lines 9, 10, and 11), and Non-Farm Income (Lines 11 through 15), to get Total Cash Available (Line 16), which then includes all sources of cash other than inflows from savings and borrowed funds. The next major section lists all Cash Operating Expenses (Lines 17 through 36), Livestock and Feed Purchases (Lines 37, 38, and 39), Capital Expenditures (Lines 40, 41, and 42), and Other Expenditures (Lines 43 through 49), to get Total Cash Required (Line 50), which then includes all uses of cash other than outflows to savings, repayments of operating loans, and cash carryover to the next period. The third major section considers all of the adjustments needed to complete the sources = uses equation and to maintain positive bank account balances. We will need to discuss each of the three major sections, but clearly we need the first two to be able to make sense of the third.

The AFRA cash flow budget provides 14 columns, one for last year (usually placed in the first column), one for the yearly total, and one for each month of the year. It also provides two additional row sections, one to maintain a monthly track of loan balances (Lines 59 through 62) and a second to allow a mathematical check on computations performed (Lines 63, 64, and 65—Line 65 should always be zero if computations were performed correctly).

Total Cash Available

Almost all of the work needed to complete the first 15 lines has already been done. The information generated previously just needs to be transferred from the schedules to the cover page. Beginning cash balance for the year (and consequently for January) comes directly from the beginning balance sheet (the business checking account only). Operating receipts must be drawn from AFRA Schedules C-4 and C-5

(Figures 7.1 and 7.3) for crops and livestock. Custom work receipts are usually estimated separately, both the total and the timing, and are usually based on last year's records or on projections of new activities for this year to more fully use labor and machinery resources during slack periods. Government payments can be forecast after consultations with the USDA Agricultural Stabilization and Conservation Service (ASCS), which will estimate amount and timing of payments based on the farm price support programs in effect for this year. Patronage dividends from cooperatives might be estimated based on last year's amounts, conditioned by this year's projections of input use and this year's marketing plans, or they might be based on the cooperative manager's forecast of coop profitability. Hedging account activity estimates are drawn directly from the marketing plans developed in Chapter 5.

Capital receipts forecasts for breeding livestock sales can be transferred from Schedule C-4 (Figure 7.1), but the receipts from machinery and equipment sales are a bit more nebulous. They tie directly to the capital replacement and purchase plans developed in Chapter 8.

Nonfarm income is relatively easy to estimate. Wages from off-farm jobs are typically much more predictable than farm income. Interest can be estimated easily if the periodic savings or bond interest rate is known. Estimates of dividends from corporate stocks should draw on the experience of a qualified stock broker.

Before the mechanical functions are performed to calculate Line 16, consider any other receipts that have not been called for specifically by the printed line titles. Are there any accounts receivable or payments on notes receivable outstanding and likely to be collected during the year? Are any gifts, inheritances, or trust payments expected during the year? Is a nonfarm business likely to contribute any cash that could be used for farm or ranch operations? Estimate all of these items or any other sources of cash and then add Lines 1 through 15 to get the total cash available in Line 16.

Total Cash Required

Most of the operating expenses projections were discussed in Chapters 6, 7, and 8 and now must be transferred from the purchasing schedules of AFRA Schedule C-3 (Figure 7.6) and the other purchasing schedules like the fuel example in Figure 7.7. Repair expense estimates can be transferred from AFRA Schedule C-6 (Figure 8.2). Storage and custom drying, as well as marketing, expenditures draw on the marketing plans developed in Chapter 5, or they can be estimated from the enterprise budgets developed in Chapter 6. The enterprise budgets also provide the forecasts, along with past farm or ranch financial records, for insurance and property taxes (see specifically the whole-farm summary in Table 6.9).

Livestock and feed purchases have already been estimated with the completion of AFRA Schedules C-4 and C-5, capital expenditures were estimated on AFRA Schedule C-6, and other expenditures were estimated on AFRA Schedules C-7 and C-8 as well as the income tax estimates shown in Figure 8.4. Again, before performing the mechanical functions to calculate Line 50, consider any other expenditures that have not been called for specifically by the printed line titles. Are there any accounts payable outstanding that must be paid during the year? Are any new nonfarm investments planned (especially consider contributions to retirement accounts or the purchase of additional stocks and bonds)? Are any gifts planned? Will significant legal

CASH FLOW BUDGET

For 19____
Date Completed: _____ , 19 _____   Name _____

1	Beginning cash balance (checking accounts plus currency)						
	Operating Receipts:						
2	Feeder livestock and poultry						
3	Crops and Feed						
4	Livestock and poultry products						
5	Custom work; cash patronage dividends						
6	Government payments (cash)						
7	Hedging account withdrawals						
8							
9	**Capital Receipts:** Breeding livestock						
10	Machinery, equipment, real estate						
11							
12	**Non-Farm Income:** Off-farm wages						
13	Interest and dividends						
14	Other businesses & investments						
15							
16	**TOTAL CASH AVAILABLE (add lines 1 thru 15)**						
	Operating Expenses:						
17	Chemicals						
18	Custom machine hire						
19	Fertilizer and lime						
20	Gas, fuel, oil						
21	Insurance (property, liability, crop)						
22	Labor hired (including all taxes and employee benefits)						
23	Livestock expenses (breeding, vet, etc.)						
24	Marketing and transportation expense						
25	Rents and leases						
26	Repairs—machinery and equipment						
27	—buildings and improvements						
28	Seed						
29	Storage and custom drying						
30	Supplies						
31	Taxes (real estate and personal property)						
32	Utilities (farm share)						
33	Auto expenses (farm share)						
34							
35							
36	**Total Cash Operating Expenses (add lines 17 thru 35)**						
37	**Livestock Purchases:** Feeder livestock						
38	Breeding livestock						
39	**Feed Purchases**						
40	**Capital Expenditures:** Machinery and equipment						
41	Buildings and improvements						
42							
	Other Expenditures:						
43	Hedging account deposits						
44	Family living withdrawals						
45	Other businesses & investments						
46	Income tax and social security						
47							
48	Term debt loan payments—principal						
49	—interest						
50	**TOTAL CASH REQUIRED (add lines 36 thru 49)**						
51	**CASH AVAILABLE LESS CASH REQUIRED (line 16 minus 50)**						
52	Inflows from savings						
53	Cash position before borrowing						
54	Money to be borrowed—operating loans						
	—term debt						
55	Operating loan payments—principal						
56	—interest						
57	Outflows to savings						
58	Ending cash balance						
	Loan Balances: (at end of period)						
59	Current year's operating loans						
60	Previous year's operating loans						
61	Term debt loans						
62	Total loans						
	Consistency Check:						
63	Total inflows including borrowed money (16 + 52 + 54)						
64	Total outflows (50 + 55 + 56 + 57 + 58)						
65	Budgeting error (63 − 64)						

This form is copyrighted. It is a violation of the U.S. Copyright Law to reproduce it in any manner. To order forms write or call Century Communications Inc. 8201 Howard St., Niles, IL 60714; 708/647-1200 or Doane Agricultural Services Co., 11701 Borman Dr., St. Louis, MO 63146; 314/569-2700.

152

Figure 9.2. AFRA cash flow budget.

Source: A.W. Oltmans, D.A. Klinefelter, and T.L. Frey, *Agricultural Financial Reporting and Analysis*, Century Communications, Inc, Niles, Ill., 1992.

								1	COMMENTS
								2	
								3	
								4	
								5	
								6	
								7	
								8	
								9	
								10	
								11	
								12	
								13	
								14	
								15	
								16	
								17	
								18	
								19	
								20	
								21	
								22	
								23	
								24	
								25	
								26	
								27	
								28	
								29	
								30	
								31	
								32	
								33	
								34	
								35	
								36	
								37	
								38	
								39	
								40	
								41	
								42	
								43	
								44	
								45	
								46	
								47	
								48	
								49	
								50	
								51	
								52	
								53	
								54	
								55	
								56	
								57	
								58	
								59	
								60	
								61	
								62	
								63	
								64	
								65	

This form is copyrighted. It is a violation of the U.S. Copyright Law to reproduce it in any manner. To order forms write or call Century Communications Inc., 6201 Howard St., Niles, IL 60714; 708/647-1200 or Doane Agricultural Services Co., 11701 Borman Dr., St. Louis, MO 63146. 314/569-2700

153

expenses be incurred with a farm ownership reorganization? Estimate all of these items or any other uses of cash, then add Lines 17 through 35 to get the subtotal for cash operating expenses on Line 36, and then add Lines 36 through 49 to get the total cash required in Line 50.

CASH FLOW MANAGEMENT

The cash flow management section, beginning with the first major calculation of cash available less cash required on Line 51 and continuing to the completion of ending cash balance on Line 58, is the heart of the process of making the cash flow budget work. It is almost inconceivable that a normal farm or ranch could perfectly match its cash inflows to its cash outflows in each month. Marketing strategies designed to maximize effective price received or to minimize effective price paid are a significant contributor to a mismatch between cash inflows and outflows, but not the most obvious reason. Any biological process of production requires inputs at the beginning of the process and a significant time lag before any output has matured sufficiently for sale. We might be able to minimize this seasonal pattern of spend, wait, and sell by selecting commodities that mature at different times throughout the year (for example, winter wheat matures in mid-summer, corn in fall, and cotton in early winter) or by storing crops for later sale, but even then it is very unlikely that cash inflows would match cash outflows exactly.

The first line of the adjustment section, cash available less cash required (Line 51), will tell us the direction that we must take. If negative, we must seek additional funds to finance the expenditures planned for that month. If positive, we can use the extra funds to repay debt, carryover to some point in the future, or increase expenditures.

Let us first address the situation where cash available less cash required is negative, which usually is the case early in the production process for any commodity or for the whole farm or ranch. If the farm or ranch business has a savings account that can be drawn upon, those funds will normally be used first to cover a cash shortfall.[1] The only other real option available on a regular basis is short-term borrowing, which will almost always carry a higher interest rate than the savings account. Thus, to minimize interest costs (short-term rate exceeds the opportunity cost of savings interest foregone), savings will be used first. The amount to transfer to the checking account (and to then enter on Line 52) depends upon three factors: amount of cash shortfall; savings account balance; and desired minimum ending cash balance level. A pre-established minimum ending cash balance (or target cash balance) is necessary to provide a cushion for incidental expenses or small budgeting errors and to provide liquidity in the early portion of the next period. This number depends on the size of the business, the business' purchasing/sales patterns within any month, bank account compensating balances, and other factors.[2] General Motors may need a contingency fund of a million dollars in cash; most family farms and ranches probably need a bit less than that, more likely closer to $1,000 or $5,000. The amount to transfer from savings to checking (Line 52) should then be the smaller of the cash shortfall plus desired ending balance, or savings balance.

If no savings remains or if it is inadequate to cover the cash shortfall plus ending balance needs, we must then turn to means other than self-financing. This usually entails borrowing short-term or operating funds from a financial institution such as a commercial bank, a Production Credit Association (or Farm Credit Services), or the Farmers Home Administration (Line 54). Short-term debt usually is available in one of three ways: a revolving line of credit; a nonrevolving line of credit; or a short-term note. Lines of credit are instruments that allow the borrower to draw funds up to a negotiated limit, as those funds are needed. A short-term note is an instrument that allows a transfer of a negotiated amount of borrowed funds directly in one transaction to the borrower's checking account. While it may seem easier to just take all of the negotiated credit in one lump sum, the line of credit is normally used because it reduces the total interest cost. In all cases, interest is charged on the outstanding daily balance. If the funds were withdrawn in one lump sum, before the entire amount was actually needed, unnecessary interest expenses would be incurred. The line of credit allows funds to be taken out only when needed, thus reducing interest costs. A revolving line of credit establishes an upper limit of the amount that can be outstanding at any particular point in time. Funds can be repaid and reborrowed a number of times, as long as the total outstanding balance is always less than the credit limit.[3] A nonrevolving line of credit establishes an upper limit of the amount that can be withdrawn. Funds that are repaid cannot be reborrowed.

It may well be that the negative cash available less cash required balance was caused by the purchase of a capital asset. If this is the case, we need to consider borrowing funds from a financial institution to finance that purchase, or we might finance it through the seller. Regardless of where we finance the purchase of the capital asset, the loan will usually be amortized or structured for repayment over a period of years. Short-term funds should seldom be used to finance the purchase of a capital asset. A longer-term note (Line 54) is a more appropriate mechanism. This new note should be entered on AFRA Schedule C-8.

Let's now turn to the opposite situation where cash available less cash required is positive, which is common after harvest. A positive funds available balance will be used in this order:

1. Maintain the desired minimum ending cash balance (Line 58)
2. Pay interest on short-term borrowed funds (Line 56)
3. Repay short-term borrowed principal (Line 55)
4. Transfer remaining excess to savings account (Line 57)

We must maintain our desired ending cash balance first and then repay our short-term note. Almost all financial institutions require that all accrued interest be repaid before any principal is repaid. Because the interest rate on savings is less than the interest rate on borrowed funds, we will place excess funds into savings last. We might deviate from this pattern, however, if the short-term funds were borrowed on a non-revolving line of credit. If there is just a temporary excess of cash available less cash required, and those borrowed funds will be necessary in a later month, we may need to delay repaying the borrowed money until we are sure that we no longer need it.

SHORT-TERM DEBT REPAYMENT

As noted in the previous section, there are three ways to borrow short-term funds from financial institutions, but there really is only one way to repay those funds: pay interest first and repay principal second from excess operating funds. The interest rate is usually stated on an annual basis and the interest cost is usually calculated on a daily basis:

$$I = \frac{i}{365} \times (\text{Days money is borrowed}) \times (\text{Amount borrowed})$$

where:

I = interest cost
i = annualized interest rate

When estimating accrued interest, separate by month the outstanding balance for that month, and calculate the interest cost for each month separately. Remember that interest must be paid first and repaying funds early under a nonrevolving line of credit should be approached cautiously.

DEBT RESTRUCTURING

In recent years, many farms and ranches have found themselves to be in severe financial difficulty because of the inability to generate the funds to meet an existing structured debt payment or the inability to repay operating funds entirely at the end of the operating year. This results in the need to carryover or rollover the unstructured loan to the next year. A cash flow planner should never plan a rollover of an operating loan; if caught in a position where insufficient income can be generated to meet all cash commitments, debt restructuring should be considered.

Debt restructuring simply involves planning to request a new debt repayment schedule (usually a longer term or a lower interest rate) for existing debt and/or new debt. Restructuring presumes that a temporarily low price or low yield year occurred and that profitable conditions are realistically expected to return very soon. Usually, all or most existing debts and any new debts are lumped together, and a repayment schedule that will allow the cash flow statement to work out is found. Mechanically, structuring the new note is no different than structuring any intermediate or long-term loan to finance the purchase of a capital asset—only the purpose is different. Furthermore, the cash flow planner must be careful to find a schedule that allows repayment and avoids problems in future years because of the farmer's or rancher's current tight financial position.[4]

REVISIONS TO THE CASH FLOW PLAN

We have now finally reached a point where we have a *complete* cash flow plan. However, it may not be the *best* cash flow plan. It may seem that we would automatically generate the best or optimal cash flow plan, because at each step in the process we maximized net income through selecting the commodities to produce, optimizing yields and input levels, maximizing output prices, and minimizing input prices. But we did not force ourselves to live within a short-term credit limitation or a new intermediate or long-term credit limitation, and we did not adequately analyze the impact of our actions on operating interest costs. Before we accept our first plan as our best, we need to remind ourselves that we have completed only the first seven of the 12-step planning process outlined in Chapter 2. To review, the steps in this iterative planning process included the following:

1. Establish short-term and long-term business performance and personal goals.
2. Prepare a balance sheet, income statement, and statement of cash flows for the previous year.
3. Determine the marketing and price outlook or projections for each of the major commodities.
4. Develop a preliminary marketing plan.
5. Determine optimal levels of inputs and output from each commodity, based upon production function principles.
6. Combine production plans for each commodity and develop an optimal whole-farm or whole-ranch plan based upon projected input and output prices and resource constraints.
7. Prepare a projected whole-farm or whole-ranch budget (or preliminary projected income statement) from the production plan prepared in Step 6.
8. Prepare a projected cash flow budget based on the plan and budget developed in Steps 6 and 7.
9. Check the levels of cash borrowing requirements projected by the cash flow budget.
    a. Are the requirements reasonable in relation to borrowing capacity?
    b. Can short-term cash flow problems be averted by modifying purchasing patterns?
    c. Can short-term cash flow problems be averted by modifying the output marketing plans?
10. Prepare a projected balance sheet (cost basis), income statement, and statement of cash flows.
11. Analyze the projected financial statements and assess projected performance.
12. Revise the marketing and production plans developed in Steps 4 and 5. Repeat the planning process beginning with Step 3 based on the new marketing plan and new restrictions on borrowing.

We must go further on this process outline and consider the questions in Step 8, and then remember that the only way to judge profitability is to complete an income statement on an accrual basis.

Most lenders have a rule of thumb or guideline for the amount of risk they are willing to assume in extending a loan. That rule or guideline will vary by geographic area and by purpose of the loan but will usually relate to two common ratios: the current ratio and the leverage ratio (or the debt-to-asset ratio). Even if the cash flow plan indicates that all new short-term debt can be repaid within the operating year with no debt restructuring, a lender may not be willing to lend the entire amount of short-term credit that is indicated by the cash flow plan. If this happens, the farmer or rancher has very little choice (other than seeking another lender) but to revise the cash flow plan to reduce borrowing needs. Normal rules of thumb (again, these rules may vary by area, lender, and/or type of operation) would be that the current ratio should never fall below 1:1 and the leverage ratio (or debt to equity ratio) should never exceed 3:1 or 4:1. It is virtually impossible to define a borrowing limit that is acceptable or a loan request that is excessively high because of the various lender-borrower factors.[5]

Revising borrowing needs downward may also work in the favor of the borrower if short-term credit needs (and thus cash interest costs) are excessively high. We must consider the impact of interest costs on net farm or ranch income, as determined with the projected income statement of Step 11.

Revisions to a cash flow plan to reduce short-term borrowing needs is an iterative or circular process as indicated by Step 12. Most of the revisions in cash flow will be made by modifying the marketing plans. For example, even though it may cost more on a per-ton basis, fertilizer expenses may have to be paid later than indicated in the first plan in order to conserve cash balances or reduce borrowing early in the growing season. Or even though a lower price will be received, it may be necessary to sell outputs closer to harvest and avoid long storage periods in order to generate the cash to repay short-term credit earlier (and thus reduce interest payments). In the extreme, it may be necessary to reduce planned expenditures (especially capital expenditures) and operate with somewhat less than optimal levels of purchased inputs in order to reduce cash needs.

Unfortunately, there is no direct procedure to follow to find the single *best* or optimal cash flow plan other than the development of an extensive matrix for the preparation of the whole-farm plan indicated in Step 6. Although finding the optimal plan is still possible if using programmed budgeting, it will likely be most efficient to develop a detailed, computerized linear programming model.

A COMPLETE CASH FLOW PLAN: SNAKE BELLY FARMS

Joe and Mary Farmer began the process of putting their complete cash flow plan together by transferring the component plans from AFRA Cash Flow Schedules C-3 through C-9 onto the overall budget form (refer to Figures 7.8 through 7.16 and 8.6 through 8.9). The second step involves estimating capital receipts, off-farm income, government payments, and investment income, as well as insurance, property taxes, and miscellaneous expenses. The final step in developing the first plan is completed by calculating lines 1, 16, 50, and 51 through 58 (Figure 9.3).

Joe and Mary estimated their total operating credit needs to be a revolving line of credit for $50,000 in addition to a cattle purchase loan of $155,000.

Joe and Mary completed their financial statement package for presentation to their lender and for analysis by transferring end-of-the year projected inventory levels from the various cash flow schedules to a new set of balance sheet schedules. They adjusted their capital inventory and depreciation schedule to reflect changes expected in 19X2 and transferred ending cash, savings, and loan balances from the cash flow statement to complete a projected balance sheet as of December 31, 19X2. Finally, they used cash receipt and expenditure estimates from the cash flow statement and inventory balances from the December 31, 19X1, and December 31, 19X2, balance sheets to prepare a projected income statement for 19X2, a projected statement of cash flows, and a projected statement of owner equity (see Figures 9.4 through 9.7).

CASH FLOW BUDGET
Snake Belly Farms
Projected for 19X2

DESCRIPTION	LAST YEAR	TOTAL	JAN	FEB	MAR	APRIL	MAY	JUNE	JULY	AUG	SEPT	OCT	NOV	DEC
BEG. CASH BALANCE	0	10000	10000	4217	1000	1000	1000	1000	8707	1000	1000	1000	18200	13270
OPERATING RECEIPTS:														
CROPS AND FEED	0	242717	6885	0	733	0	0	0	0	9338	0	91213	0	134548
LIVESTOCK & POULTRY	0	247752	0	0	0	0	0	247752	0	0	0	0	0	0
PRODUCTS (LIVESTOCK)	0	0	0	0	0	0	0	0	0	0	0	0	0	0
CUSTOM WORK	0	0	0	0	0	0	0	0	0	0	0	0	0	0
GOVERNMENT PAYMENTS	0	23392	0	0	0	6109	0	0	4872	0	3248	0	0	9163
HEDGING ACCOUNT W/D	0	0	0	0	0	0	0	0	0	0	0	0	0	0
CAPITAL RECEIPTS:														
BREEDING STOCK	0	0	0	0	0	0	0	0	0	0	0	0	0	0
MACHINERY & EQUIP.	0	0	0	0	0	0	0	0	0	0	0	0	0	0
NON-FARM INCOME:														
OFF-FARM WAGES	0	14400	1200	1200	1200	1200	1200	1200	1200	1200	1200	1200	1200	1200
INTEREST AND DIVIDENDS	0	0	0	0	0	0	0	0	0	0	0	0	0	0
TOTAL CASH AVAILABLE	0	538261	18085	5417	2933	8309	2200	249952	14779	11538	5448	93413	19400	158181
OPERATING EXPENSES:														
LABOR HIRED	0	12900	850	850	850	850	850	2200	850	850	850	850	850	2200
REPAIRS-MACH & EQUIP.	0	9120	760	760	760	760	760	760	760	760	760	760	760	760
REPAIRS-BUILD/IMPROV.	0	2578	75	75	75	75	75	1075	753	75	75	75	75	75
RENTS & LEASES	0	0	0	0	0	0	0	0	0	0	0	0	0	0
SEED	0	8591	0	0	0	600	2431	0	0	5560	0	0	0	0
FERTILIZER & LIME	0	18807	0	0	0	1350	0	0	0	17457	0	0	0	0
CHEMICALS	0	12849	0	0	0	6400	675	1920	2625	533	696	0	0	0
CUSTOM MACHINE HIRE	0	9387	0	0	0	0	0	3031	0	0	0	0	0	6356
SUPPLIES	0	1500	125	125	125	125	125	125	125	125	125	125	125	125
LIVESTOCK EXPENSE	0	5400	780	600	600	600	600	600	0	0	0	420	600	600
GAS, FUEL, OIL	0	58018	0	0	0	7891	8216	8704	9660	10930	10406	511	1700	0
STORAGE/CUSTOM DRY	0	0	0	0	0	0	0	0	0	0	0	0	0	0
TAXES (REAL EST. PP)	0	2112	0	2112	0	0	0	0	0	0	0	0	0	0
INSURANCE(PROP,LIAB)	0	3300	800	0	1000	1500	0	0	0	0	0	0	0	0
UTILITIES(ELECT/GAS)	0	2040	170	170	170	170	170	170	170	170	170	170	170	170
MARKET/TRANSPORT EXP	0	0	0	0	0	0	0	0	0	0	0	0	0	0
AUTO (FARM SHARE)	0	4200	350	350	350	350	350	350	350	350	350	350	350	350
ACCOUNTS PAYABLE	0	7000	7000	0	0	0	0	0	0	0	0	0	0	0
EDUC, TRG, & MISC.	0	4983	0	200	1233	350	1200	800	0	1200	0	0	0	0
TOTAL CASH OPER EXPS	0	162785	10910	5242	5163	21021	15452	19735	15293	38010	13432	3261	4630	10636

Figure 9.3. A complete cash flow plan, Snake Belly Farms.

CASH FLOW BUDGET
Snake Belly Farms
Projected for 19X2

DESCRIPTION	LAST YEAR	TOTAL	JAN	FEB	MARCH	APRIL	MAY	JUNE	JULY	AUG	SEPT	OCT	NOV	DEC
STOCK & FEED PURCH:														
FEEDER LIVESTOCK	0	173250	0	0	0	0	0	0	0	0	0	17325	0	0
BREEDING LIVESTOCK	0	0	0	0	0	0	0	0	0	0	0	0	0	0
FEED PURCHASED	0	4594	1458	0	0	0	0	0	0	0	0	3136	0	0
CAPITAL EXPENDITURES														
MACHINERY & EQUIP	0	17000	0	0	2000	0	0	0	0	0	0	15000	0	0
BUILDINGS & IMPROVE	0	3710	0	0	0	0	0	0	3710	0	0	0	0	0
OTHER EXPENDITURES:														
HEDGING ACCT DEPOSIT	0	0	0	0	0	0	0	0	0	0	0	0	0	0
GROSS FAMILY LIV W/D	0	18000	1500	1500	1500	1500	1500	1500	1500	1500	1500	1500	1500	1500
NON-FARM BUS/INVEST	0	0	0	0	0	0	0	0	0	0	0	0	0	0
INCOME TAX & SOC SEC	0	1000	0	0	1000	0	0	0	0	0	0	0	0	0
	0	0	0	0	0	0	0	0	0	0	0	0	0	0
LOAN PAYMENTS - PRIN	0	206433	0	0	0	0	0	165000	0	0	0	0	0	41433
LOAN PAYMENTS - INT	0	104512	0	0	0	0	0	13200	0	0	0	0	0	91312
TOTAL CASH REQUIRED	0	691284	13868	6742	9663	22521	16952	199435	20503	39510	14932	19617	6130	144881
CASH AVAIL - CASH REQ		-153023	4217	-1325	-6730	-14212	-14752	50517	-5724	-27972	-9484	-10274	13270	13300
INFLOWS FROM SAVINGS		0	0	0	0	0	0	0	0	0	0	0	0	0
CASH POS BEFORE BORR		-153023	4217	-1325	-6730	-14212	-14752	50517	-5724	-27972	-9484	-10274	13270	13300
MONEY TO BE BORROWED														
-OPERATING LOANS		87576	0	2325	7730	15212	16129	0	6724	28972	10484	16800	0	0
-INT & L/T LOANS		168000	0	0	0	0	0	0	0	0	0	168000	0	0
OP LOAN PAY - PRIN		87576	0	0	0	0	0	41396	0	0	0	46380	0	0
-INTEREST		1677	0	0	0	0	377	414	0	0	0	186	0	0
OUTFLOWS TO SAVINGS		0	0	0	0	0	0	0	0	0	0	0	0	0
ENDING CASH BALANCE		13300	4217	1000	1000	1000	1000	8707	1000	1000	1000	18300	13270	13300
LOAN BALANCES:														
CURR INTEREST RATE	12.00													
CURRENT YR'S OP LOAN			0	2325	10055	25267	41396	0	6724	33696	46180	0	0	0
-ACCRUED INTEREST		1677	0	23	101	253	414	0	67	357	462	0	0	0
PREV YR'S OPER LOANS			0	0	0	0	0	0	0	0	0	0	0	0
-ACCRUED INTEREST			0	0	0	0	0	0	0	0	0	0	0	0
INT & LONG TERM LOAN	1151995		1151995	1151995	1151995	1151995	1151995	986995	986995	986995	986995	1154995	1154995	1113562
TOTAL LOANS			1151995	1151995	1151995	1151995	1151995	986995	993719	1022691	1033175	1154995	1154995	1113562
CONSISTENCY CHECK:														
TOTAL INFLOWS		18085	18085	7742	10663	23521	18329	249952	21503	40510	15932	26813	19400	158181
TOTAL OUTFLOWS		18085	18085	7742	10663	23521	18329	249952	21503	40510	15932	26813	19400	158181
BUDGETING ERROR		0	0	0	0	0	0	0	0	0	0	0	0	0

Figure 9.3. (continued).

BALANCE SHEET
Snake Belly Farms
December 31, 19X2

ASSETS			LIABILITIES		
CURRENT ASSETS	COST	MARKET VALUE	**CURRENT LIABILITIES**		MARKET VALUE
Cash on hand and in checking accounts	13,303	13,303	Notes payable		155,000
Feeder livestock & poultry	192,456	192,456	Principal due within 12 months		
Crops and feed	8,550	8,550	on all non-current liabilities		48,695
Investment in growing crops	40,800	40,800	Accrued interest on:		
Supplies	2,180	2,180	Accounts 0 Notes 4,623		
			Non-current liabilities 4,314		8,937
			Accrued tax liability:		
			Real estate		1,625
TOTAL CURRENT ASSETS	257,289	257,289	TOTAL CURRENT LIABILITIES		214,257
NON-CURRENT ASSETS			**NON-CURRENT LIABILITIES**		
Machinery, equipment, trucks		163,475	(Principal due beyond 12 months)		
Cost or basis	383,950		Notes payable		909,867
Less accumulated dep	356,014	27,936			
Real estate		914,000			
Cost or basis	1,000,000		TOTAL NON-CURRENT LIABILITIES		909,867
Less accumulated dep	66,000	934,000			

				COST	MARKET VALUE
			TOTAL CURRENT AND NON-CURRENT LIABILITIES		
TOTAL NON-CURRENT ASSETS	961,936	1,088,975		1,124,124	1,124,124
TOTAL BUSINESS ASSETS	1,219,225	1,346,264			
			Deferred tax on non-current assets		53,302
Personal Assets		21,900			
			TOTAL LIABILITIES	1,124,124	1,177,426
			Owner equity:		
			Retained Earnings	75,101	75,101
			Contributed Capital	20,000	20,000
			Personal Net Worth	--	21,900
			Valuation Equity	--	73,737
TOTAL ASSETS	1,219,225	1,368,164	TOTAL OWNER EQUITY	95,101	190,738
			TOTAL LIABILITIES & OWNER EQUITY	1,219,225	1,368,164

A. December 31, 19X2, projected balance sheet.

DESCRIPTION	QUANTITY	UNIT	VALUE
BALING TWINE	40	ROLLS	1,000
DIESEL FUEL	330	GALS	330
MOTOR OIL	12	GALS	50
MISCELLANEOUS			800
		TOTAL	2,180

B. Supplies schedule.

Figure 9.4. Projected balance sheet, Snake Belly Farms, December 31, 19X2.

	% OWN	DATE ACQ	COST OR BASIS	ACC DEP	ADJUSTED VALUE	MARKET VALUE
LAND (400 ACRES, CLASS I)	100	12/86	440,000	0	440,000	400,000
RESIDENCE	100	12/86	50,000	0	50,000	55,000
SERVICE BLDG	100	12/86	90,000	18,000	72,000	80,000
IMPROVEMENTS	100	12/86	160,000	32,000	128,000	150,000
LAND (200 ACRES, CLASS II)	100	12/86	120,000	0	120,000	100,000
IMPROVEMENTS	100	12/86	80,000	12,000	68,000	75,000
LAND (120 ACRES, CLASS III)	100	12/86	60,000	0	60,000	54,000
ACRES 720		TOTAL TRACT	1,000,000	66,000	934,000	914,000

C. Real estate schedule.

TO WHOM	PURPOSE OR SECURITY	DUE DATE	INT RATE	PAY DATE	CURR PRIN BALANCE	PART DUE (12 MOS)	PRIN DUE (12 MOS)	ACCRUED INT
1ST NATL	19X3 FORD	10/01/X4	11.00	10/01	13,000	2,760	10,240	355
1ST NATL	TON BALER	12/01/X4	14.00	12/01	31,176	14,569	16,607	357
1ST NATL	3 PUMPS	12/15/X3	13.00	12/15	3,774	3,774	0	21
PREVIOUS OWNER	FARM	12/15/Y4	9.00	12/15	910,612	27,592	883,020	3,581
TOTALS					958,562	48,695	909,867	4,314

D. Noncurrent debt schedule.

ITEM	YEAR, MAKE AND MODEL	DATE ACQ	% OWN	COST OR BASIS	ACC DEPR	ADJUSTED COST	MARKET VALUE
65 HP TRACTOR		1963	100	5,500	5,500	0	1,500
96 HP TRACTOR		1971	100	10,300	10,300	0	4,100
130 HP TRACTOR		1984	100	43,600	43,600	0	27,000
2-ROW COTTON PICKER		1985	100	77,000	77,000	0	26,500
14 FT. SWATHER		1986	100	29,000	29,000	0	9,000
1-TON BALER		1987	100	65,000	41,350	23,650	27,000
4-ROW PLANTER		1985	100	6,800	6,800	0	3,400
4-ROW CULTIVATOR		1983	100	3,000	3,000	0	1,500
14 FT. OFFSET DISK		1983	100	11,500	11,500	0	5,750
13 FT. DRILL		1985	100	4,300	4,300	0	2,150
12 FT. PLANE		1983	100	5,000	5,000	0	2,500
14 FT. FLOAT		1983	100	900	900	0	600
4-ROW LISTER		1983	100	3,200	3,200	0	1,600
4-16 IN. MOLDBOARD PLOW		1983	100	8,000	8,000	0	4,000
4-ROW SHREDDER		1983	100	5,000	5,000	0	2,500
8 COTTON TRAILERS		1980	100	19,200	19,200	0	9,600
12 FT. SPRAYER		1983	100	2,750	2,750	0	1,375
FRONT-END LOADER (ATTACH)		1983	100	5,100	5,100	0	2,550
SEMI-TRUCK & FLATBED		1985	100	30,000	30,000	0	15,000
RAKES		1983	100	6,800	6,800	0	3,400
V-DITCHER		1983	100	2,500	2,500	0	1,250
3 NATURAL GAS PUMPS		1986	100	15,000	15,000	0	7,500
PICKUP	1978 FORD	1978	100	7,000	7,000	0	800
PICKUP	1971 FORD	1971	100	2,500	2,500	0	400
PICKUP	1991 FORD	1990	100	15,000	10,714	4,286	14,000
		TOTALS		383,950	356,014	27,936	174,975

E. Machinery schedule.

Figure 9.4. (continued).

ITEM	QUANTITY	UNIT	$/UNIT	VALUE
SOYBEANS	0		5.36	0
CORN	0		2.15	0
ALFALFA	100		85.50	8,550
			TOTAL	8,550

F. Crops and feed schedule.

DESCRIPTION	NUMBER	AVERAGE WEIGHT	$/CWT	VALUE
STOCKER STEERS	540	540	66.00	192,456
			TOTAL	192,456

G. Feeder livestock schedule.

CROP	ACRES	$/ACRE	VALUE
ALFALFA I	300	100.00	30,000
PASTURE II	100	50.00	5,000
PASTURE III	100	40.00	4,000
ALFALFA II	20	90.00	1,800
		TOTAL	40,800

H. Investment in growing crops schedule.

TO WHOM	PURPOSE OR SECURITY	DUE DATE	INT RATE	PAY DATE	PRINCIPAL BALANCE	ACCRUED INT
1ST NATL	CATTLE	06/01/X3	12.00	10/01	155,000	4,623
			TOTAL		155,000	4,623

I. Notes due within 12-months schedule.

DESCRIPTION	VALUE
SUBARU (1981 GL)	1,900
MG (1979 B)	3,000
HOUSEHOLD GOODS UNIVERSAL LIFE	12,000
INSURANCE POLICY (MUTUAL; 100,000 FACE)	5,000

J. Personal vehicles and household goods schedule.

Figure 9.4. (continued).

Section A - Deferred Tax Estimate on Current Assets

Feeder livestock and poultry	192,456		
(minus) Purchase cost of feeders on hand	-170,100	22,356	
Crops and feed		8,550	
Supplies		2,180	
Investment in growing crops		40,800	
Total current assets that could be taxed			73,886
Estimated accrued interest		8,937	
Estimated accrued tax liability		1,625	
Total current liabilities that could be deducted			-10,562
NET TAXABLE CURRENT ASSETS			63,324

Section B - Deferred Tax Estimate on Non-Current Assets

	Cost		Market Value		Taxable Gain		
Machinery	-	27,936	+	174,975	+	147,039	
Farm real estate	-	934,000	+	914,000	-	20,000	
NET TAXABLE GAINS						127,039	
NET TAXABLE INCOME							190,363
Ordinary tax rate						X	28%
DEFERRED TAX							53,302

K. Deferred tax liability worksheet.

Figure 9.4. (continued).

INCOME STATEMENT
Snake Belly Farms
Projected for 19X2

REVENUE

Feeder Livestock & poultry:

Cash sales	247,752		
Inventory change	0		
Feeder livestock transferred to breeding herd	0	247,752	

Crops and feed:

Cash sales	242,717		
Inventory change	-4,250	238,467	
Government payments		23,392	
Gross revenue			509,611
minus feeder livestock & poultry purchases		173,250	
minus feed purchased		4,594	
VALUE OF FARM PRODUCTION			331,767

EXPENSES

Cash operating expenses		162,785	
Accrual expense adjustments (unused assets & unpaid items)		-6,785	
Depreciation: Machinery and equipment	23,604		
Fixed farm improvements	16,500	40,104	
Total operating expenses			196,084
Income from farm operations			135,683
minus Interest expense			105,956
NET FARM INCOME FROM OPERATIONS			29,727
Capital adjusment, gain (loss)		0	
NET FARM INCOME			29,727

NON FARM INCOME

Spouse's wage off farm		14,400	
NON-FARM INCOME			14,400
INCOME BEFORE INCOME TAXES AND EXTRAORDINARY ITEMS			44,127
Income and social security taxes			1,000
Income before extraordinary items			43,127
Extraordinary items (explain)			0
NET INCOME			43,127

A. 19X2 Projected income statement.

Figure 9.5. 19X2 Projected income statement, Snake Belly Farms.

	BEG INVENTORY	END INVENTORY	CHANGE
Accounts receivable	- 0	+ 0	0
Livestock & poultry to be sold	- 192,456	+ 192,456	0
Crops and feed	- 12,800	+ 8,550	-4250

B. Accrual adjustments to revenue schedule.

ITEM (excluding interest, feed, livestock)	CASH EXPENSE
Labor (including taxes and benefits)	12,900
Repairs - machinery and equipment	9,120
Repairs - building and improvements	2,578
Seed	8,591
Fertilizer and lime	18,807
Chemicals	12,849
Custom machine hire	9,387
Supplies	1,500
Livestock expenses	5,400
Gas, fuel, oil	58,018
Taxes (real estate/personal property)	2,112
Insurance (property, liability, crop)	3,300
Utilities (farm share)	2,040
Auto (farm share)	4,200
Education, training, and miscellaneous	4,983
19X1 accounts payable	7,000
TOTAL	162,785

C. Cash operating expenses schedule.

	BEG INVENTORY	END INVENTORY	CHANGE
Unused assets			
Cash investment in growing crops	+ 40,800	- 40,800	0
Supplies	+ 0	- 0	0
Prepaid expenses	+ 2,375	- 2,180	195
TOTAL	+ 43,175	- 42,980	195
Unpaid items			
Farm accounts payable	- 7,000	+ 0	-7,000
Accrued real estate taxes	- 1,625	+ 1,625	0
TOTAL	- 8,625	+ 1,625	-7,000

D. Accrual adjustments to expenses schedule.

Amount paid in cash or by renewal		106,189
Accrued interest		
Beginning of period	- 9,170	
End of period	+ 8,937	
Adjustment for change in accrued interest		- 233
TOTAL INTEREST EXPENSE		105,956

E. Interest expense schedule.

Figure 9.5. (continued).

STATEMENT OF CASH FLOWS
Snake Belly Farms
Projected 19X2

CASH FLOWS FROM OPERATING ACTIVITIES:

Cash received from farm operations:			
Feeder livestock and poultry sales	247,752		
Crops and feed	242,717		
Government payments, cash and certificates	23,392	513,861	
Cash received from non-farm income and operations:			
Wages		14,400	
Cash paid for farm operating activities:			
Feeder livestock and poultry	173,250		
Feed purchases	4,594		
Interest expense	106,189		
Operating expenses	162,785	-446,818	
Income and Social Security Taxes		-1,000	
NET CASH INCOME			80,443
Cash withdrawals for family living		-18,000	
NET CASH PROVIDED BY OPERATING ACTIVITIES			62,443

CASH FLOWS FROM INVESTING ACTIVITIES:

Cash paid to purchase:			
Machinery and equipment		17,000	
Farm real estate: other farm assets		3,710	
NET CASH PROVIDED BY INVESTING ACTIVITIES			-20,710

CASH FLOWS FROM FINANCING ACTIVITIES

Operating and CCC loans received (including interest paid by loan renewal)		87,576	
Term debt financing-loans received		168,000	
Operating debt principal payments		-87,576	
Term debt principal payments: Scheduled payments		-206,433	
NET CASH PROVIDED BY FINANCING ACTIVITIES			-38,433
NET INCREASE (DECREASE) IN CASH AND CASH EQUIVALENTS			3,300

Cash and cash equivalents reported on the beginning-of-year balance sheet:	10,000
Cash and cash equivalents, as calculated, at the end of year	13,300

Figure 9.6. 19X2 Projected statement of cash flows, Snake Belly Farms.

STATEMENT OF OWNER EQUITY
SNAKE BELLY FARMS
Projected for 19X2

			Cost	Market Value
TOTAL OWNER EQUITY, Beginning of period			75,681	145,344
Change in contributed capital and retained earnings:				
Net income (loss) after taxes for the period		43,127		
Withdrawals of net income and retained earnings (cash or property) during the period:				
Withdrawals for family living expenses	23,707			
Withdrawals for investments into personal assets	0	-23,707		
Additions of capital (cash or property) to the business during the period:				
Gifts and inheritances received; additions to paid-in-capital	0			
Investments of personal assets into the business	0	0		
Distributions of capital, dividends, or gifts made (cash or property) during the period		0		
TOTAL CHANGE IN CONTRIBUTED CAPITAL AND RETAINED EARNINGS			19,420	19,420
Change in personal net worth:		21,900		
Personal net worth, end of period		-21,500		
Personal net worth, beginning of period				
TOTAL CHANGE IN PERSONAL NET WORTH			xx	400
Change in valuation equity:				
Valuation equity, end of period		73,737		
Valuation equity, beginning of period		-48,163		
TOTAL CHANGE IN VALUATION EQUITY			xx	25,574
TOTAL OWNER EQUITY, End of period			95,101	190,738

Figure 9.7. 19X2 Projected statement of owner equity.

SUMMARY

Completion of the first or preliminary cash flow plan begins with the transfer of previously projected receipts and expenditures from the AFRA cash flow schedules to the overall cash flow budget. The process continues with estimates or projections of items not included on the schedules and climaxes with the cash deficit/surplus adjustment process, which involves using and replenishing savings, borrowing new intermediate or long-term funds, debt restructuring if necessary, and borrowing and repaying short-term or operating funds.

Operating credit may be extended through a short-term note, a nonrevolving line of credit, or a revolving line of credit. Regardless of the repayment terms under which funds are borrowed, interest is repaid in full before any principal is repaid. Only if all other obligations have been met should excess funds be transferred to an interest-bearing savings account. When projecting credit needs, all short-term funds borrowed should have a compensating repayment plan within the period. If this cannot be accomplished, the planner should investigate the restructuring of existing debt to reduce current principal payments. Debt restructuring should be approached cautiously, however. Substituting debt for income is dangerous in the short run and disastrous in the long run.

The first complete cash flow plan may not be acceptable to the lender or the farm or ranch operator. Revisions, following the general 12-step process, may be necessary to reduce borrowing needs or interest payments. The effects of each revision should be viewed not only through the cash flow budget but also through the projected balance sheet and projected income statement.

NOTES

1. It is usually recommended that a family farm or ranch maintain at least four bank accounts: business checking and savings and personal checking and savings. See J.D. Libbin, *Farm and Ranch Bank Accounts*, Guide Z-405, Cooperative Extension Service, New Mexico State University, Las Cruces, September 1984.

2. An excellent discussion of cash management strategies in a corporate environment and mathematical models for determination of the optimal target cash balance is found in E.F. Brigham and L.C. Gapenski, *Financial Management: Theory and Practice*, 6th ed., Dryden Press, Chicago, Ill., 1991.

3. The typical example of a revolving line of credit is a personal credit card.

4. Additional discussion of debt restructuring and other business restructuring issues may be found in D.A. Klinefelter, *Restructuring the Farm Business*, B-1549, Texas Agricultural Extension Service, Texas A&M University, College Station, May 1989.

5. For a further discussion of borrower-lender relationships, see R.P. Sullivan and J.D. Libbin, *Evaluation of Agricultural Loan Applications*, Guide Z-703, Cooperative Extension Service, New Mexico State University, Las Cruces, August 1984, and T.L. Frey and R.H. Behrens, *Lending to Agricultural Enterprises*, Bankers Publishing Company, Boston, Mass., 1981.

RECOMMENDED READINGS

James D. Libbin, *Farm and Ranch Bank Accounts*, Guide Z-405, Cooperative Extension Service, New Mexico State University, Las Cruces, September 1984.

R. Patrick Sullivan and James D. Libbin, *Evaluation of Agricultural Loan Applications*, Guide Z-703, Cooperative Extension Service, New Mexico State University, Las Cruces, August 1984.

Royce A. Hinton, Delmar F. Wilken, and Charles E. Cagley, "Adjustments to Meet Cash Flow Needs," *Farm Economics Facts and Opinions*, 85-9, Department of Agricultural Economics, University of Illinois, Urbana, July 1985.

Royce A. Hinton, Delmar F. Wilken, and Charles E. Cagley, "Projecting Income and Cash Flow Needs—Basis for Farm Adjustments," *Farm Economics Facts and Opinions*, 85-8, Department of Agricultural Economics, University of Illinois, Urbana, July 1985.

James D. Libbin, *Calculating Interest Charges on Loans*, Guide Z-701, Cooperative Extension Service, New Mexico State University, Las Cruces, September 1984.

James D. Libbin, *Long-Term Loan Repayment Methods*, Guide Z-702, Cooperative Extension Service, New Mexico State University, Las Cruces, October 1984.

Billy B. Rice, "Cash Flow Projection for Operating Loan Determination," *Farm Management Planning Guide*, Section VIII, No. 1–A, Cooperative Extension Service, North Dakota State University, Fargo, N.Dak., September 1980.

Danny A. Klinefelter, *Restructuring the Farm Business*, B-1549, Texas Agricultural Extension Service, Texas A&M University, College Station, May 1989.

Eugene F. Brigham and Louis C. Gapenski, *Financial Management: Theory and Practice*, 6th ed., Dryden Press, Ft. Worth, Tex., 1991.

CHAPTER 10

Assessing Risk through Cash Flow

Risk is a fact of life in agriculture. Risk has become a common word wherever farmers and ranchers gather. But what is risk? And how do we measure it? Oddly enough, these two difficult questions are still subjects of major research efforts by agricultural economists. Risk can be simply defined as the possibility that some loss might occur. In the normal sense, this means the loss of profitability, but it may also mean, in the extreme, the loss of life or the loss of the entire farm or ranch.

We have some statistical measures to appeal to in attempting to measure risk, but at this time those statistics are debatable as to whether they really provide us with much information or even correct information. What we really want to know includes the following:

1. What are the causes of risk?
2. What are the effects of risk on our business?
3. How can we control it?

To control risk, we must begin with the causes and then attempt to assess its effects. But before we jump in, we must emphasize that risk is not all bad.

Microeconomics clearly tells us that if excess profits exist in any competitive industry (defined as positive economic returns to risk), the existing producers in that industry will expand production and/or new producers will enter the industry. In either case, the industry supply curve will be pushed to the right, equilibrium price will fall, and excess profits will vanish. This theory, however, assumes a deterministic world, one in which there is no risk or variability in either price or production. In a world in which there is risk, excess profits may exist, if those profits are viewed as being insufficient to outweigh the risk of loss from expansion or entrance in the industry. The net result of these elements of economic theory is that if no risk exists in the industry, then there can be no profit.

A widely accepted list of the sources of risk in agriculture includes the following seven different types:

1. Production and yield
2. Market and price
3. Business and financial
4. Technology and obsolescence
5. Casualty loss
6. Social and legal
7. Human

Although these risks are readily apparent to some, definitions might be useful to clarify the sources of risk. Nelson, Casler, and Walker[1] define these seven sources of risk as:

1. Production and yield risk. This source of risk is due to the variability in yields and production caused by such unpredictable factors as weather, disease, pests, genetic variations, and timing of practices. Examples include variations in crop yields, animal weaning weights, product quality, animal rate of gain, pasture carrying capacity, feed conversion, death loss, labor required, machinery breakdown, etc.

2. Market and price risk. This refers to the variability and unpredictability of prices that farmers receive for products and pay for production inputs. Types of variation in prices that are relatively predictable are trends, commodity cycles, and seasonal variation. In addition, random price variations result from changing supply and demand conditions, including such influences as buyer and seller expectations, speculation, government programs, and consumer demand.

3. Business and financial risk. This source of risk relates to the financing of the business, the assets it controls, and its credit obligations. This type of risk has become more important with the larger capital investments required in agriculture today as well as the increased use of borrowed capital. Variable cash flows increase the risk of not having adequate cash to meet debt payments and other financial obligations. Another example of this source of risk is the possibility of losing the lease on the land being farmed.

4. Technology and obsolescence risk. The rapid development of new technology can make current production methods obsolete shortly after important investments have been made. Adopting new technologies too soon or too late is a risk farmers must face. For example, when planning to purchase a new tractor, a farmer should consider the risk that technological advances could result in a more efficient tractor that will make the one purchased obsolete within a short time.

5. Casualty loss risk. This is a traditional source of risk referring to the loss of assets to fire, wind, hail, flood, and theft. Although it's not a new source of risk, inflation has greatly increased the value of potential losses.

6. Social and legal risk. Governmental laws and regulations are a major source of uncertainty for farmers. Examples include environmental protection; controls on the use of feed additives, insecticides, and herbicides; and land-use planning. These stem from changing social attitudes. In addition there is also the risk of law suits from liabilities due to such things as farm accidents and misuse of chemicals.

7. Human risk. The character, health, and behavior of individuals are unpredictable and contribute to the risk in farm management. The possibility of losing a key employee during a critical production period is one example of this risk. Dishonesty and undependability of business associates are others. The disabling of the farm manager can be very disruptive to the continuity of an efficient farm operation. Also, family needs and goals change, sometimes unpredictably.

In the context of normal business operations, price and production risks are the most important sources because they are the most likely to affect us on a recurring basis. The effects of these two types of risk are most immediately felt in our cash flow position, and in fact, it is the cash flow budget that best shows the short-run impact of risk on our operating plans. Before we commit fully to an operations plan, we should attempt to define and assess the risks that we are taking.

SOURCES OF PRICE AND PRODUCTION RISK

Price and production risks are common and have four major sources:

1. Increases in input prices
2. Decreases in output prices
3. Decreases in saleable yield
4. Increases in non-yield-affecting costs

Input and output markets in agriculture are characterized by highly variable prices, primarily because the supply of and demand for each commodity can only be estimated with significant error. This error is caused by fluctuating weather and pest conditions, many small producers with poor estimation capabilities of their final output levels, and fickle consumer tastes and preferences. Contracted, effective input and output prices to farmers or ranchers can change dramatically, sometimes in their favor and sometimes against. Furthermore, weather and pest conditions are virtually impossible to predict accurately over the growing season. Similarly, critical machinery breakdowns or custom operator shortages can occur at any time (according to Murphy's Law, they usually occur at the worst possible time), thereby increasing production costs unexpectedly.

RISK-RATED MANAGEMENT STRATEGIES

An excellent, easily applicable method of viewing the impact of price and production risks is the risk-rated management strategies system developed by agricultural economists John Ikerd and Kim Anderson.[2,3] Risk-rated management concepts begin with a delineation of five probabilistic outcomes: best, optimistic, expected, pessimistic, and worst. They define the expected outcome as the one outcome (e.g., price, yield, or cost of production) most likely to occur. This is the same outcome or assumption that we have used throughout the first nine chapters of this book. The optimistic outcome, like the expected outcome, is based on the planner's best estimate of what might happen; it is an amount subjectively estimated by the planner that a particularly favorable outcome (price, yield, or cost of production again) will occur with the probability of one-sixth (or will occur in one year out of six). The pessimistic outcome is exactly opposite to the optimistic—a one-sixth chance that a particularly unfavorable outcome will occur. Thus, there is a two-thirds probability (roughly one standard deviation from the mean) that the outcome will fall somewhere between the pessimistic and optimistic levels. That outcome is characterized by the expected level. The best and worst levels, often the starting points in a process of eliciting subjective price and/or yield probability distributions from farm and ranch managers, reflect a one year in 50 likelihood. This one-fiftieth level roughly conforms to the best (or worst) likely outcome in the manager's productive career. Statistically, there is an approximately 95 percent chance (or about two standard deviations from the mean) that the actual outcome will fall between the best and worst levels. Eliciting these two outer boundaries helps avoid an anchoring bias (i.e., a tendency to specify outcomes within a very narrow range).

Ikerd and Anderson often use the example of rolling two normal six-sided dice to teach concepts of probability and risk-rated management strategies. Rolling two six-sided dice will yield 36 different possible combinations (see Figure 10.1). The first die will result in any number from 1 to 6 as will the second die. Summing the results of the two dice will provide a number from 2 to 12. The outcome 2 will occur only

Possible Combinations

Die #1	Die #2			Outcomes Die #1					
1	1			1	2	3	4	5	6
1	2								
1	3		1	2	3	4	5	6	7
1	4								
1	5	D	2	3	4	5	6	7	8
1	6								
2	1	I	3	4	5	6	7	8	9
2	2								
2	3	E	4	5	6	7	8	9	10
2	4								
2	5	#	5	6	7	8	9	10	11
2	6								
3	1	2	6	7	8	9	10	11	12
3	2								
3	3								
3	4								
3	5			Outcome			Chances		
3	6			2			1/36		
:	:			3			2/36		
:	:			4			3/36		
6	1			5			4/36		
6	2			6			5/36		
6	3			7			6/36		
6	4			8			5/36		
6	5			9			4/36		
6	6			10			3/36		
				11			2/36		
				12			1/36		
							36/36		

Figure 10.1. Outcomes from rolling two six-sided dice.

with a probability of 1/36 as there is only one combination (1–1) that will sum to 2. The outcome 3, however, will occur with a probability of 2/36 since there are two combinations (1–2 and 2–1) that will sum to 3. And similarly, the outcome 4 will occur with a probability of 3/36 since there are three combinations (1–3, 2–2, and 3–1) that will sum to 4. Let us then relate the pessimistic outcome of the previous paragraph to an outcome of 2, 3, or 4. The probability of getting the outcome 2, 3, or 4 is 1/6 (1/36 + 2/36 + 3/36 = 6/36 = 1/6). We would make the same arguments for the outcomes 10, 11, or 12, which we will relate to the optimistic outcome, since these three outcomes also combine to generate a probability of 1/6.

An excellent learning exercise is to take an example like the following, place your own bets (or act as if your own money is on the line), and roll the dice to see the effect on your own net worth.[4] To put all of this in a clearer light, consider the stocker steer example provided by Anderson and Ikerd[5] (Figure 10.2).

Stocker Cattle:	BASE
PRICE	
Optimistic:	$ 74.00
Most Likely:	67.00
Pessimistic:	57.00
SALE WEIGHT: (cwt.)	
Optimistic	7.20
Most Likely:	6.25
Pessimistic:	6.25
Cost per head	$440.00
Correlation P-W	0.00
Production (hd)	100
FINANCING:	
Owner:	$ 15,000
Borrowed:	$ 29,000
Interest Rate (%)	14
NET RETURNS TO RISK: (PER HEAD)	
Optimistic:	$65.70
Most Likely:	$12.25
Pessimistic:	-$60.04
PROFIT PROB. (0%)	57
FINANCIAL RISK: TOTAL RETURNS TO EQUITY	
Optimistic:	$7620
% Equity:	51
Most Likely:	$2275
% Equity:	15
Pessimistic:	-$4954
% Equity:	-33
TARGET RETURN LEVEL:	
Prob. = > $5,000 31%	

Figure 10.2. Stocker steer risk management example.

Source: K.B. Anderson and J.E. Ikerd, *Risk-Rated Management Strategies for Farm and Ranch Decisions,* Extension Circular E-841, Cooperative Extension Service, Oklahoma State University, Stillwater, Okla., 1984.

Market risk management must be integrated with management of production and financial risk in developing meaningful risk management strategies. Market risks, production risks, and financial risks all can be included in a budget illustration utilizing the risk-rated approach. The example shown here represents a stocker cattle enterprise. The expected or most likely price is $67 in the fall for cattle to be sold the following spring. There is an estimated one-in-six chance of a price of $74 or higher and a one-in-six chance of a price of $57 or lower. The optimistic price is 10% above the expected price; the pessimistic price is 15% below. This is not inconsistent with accuracy of past price forecasts of competent market analysts. The decision maker is just being a little more conservative on the optimistic side. Expected, optimistic, and pessimistic break-even production yields, likewise, are believed to be realistic, although each producer faces a different production risk situation. No correlation is assumed to exist between prices and yields. One hundred head of cattle are to be produced. The producer is putting up $15,000, $150 per head, of his or her own money and is borrowing $29,000. An opportunity interest rate of 14 percent is assumed for the producer's equity capital. Estimated net returns reflected an expected or most likely return to risk of $12 per head. However, there is an estimated one-in-six chance of a $66 plus per head profit and an equal one-in-six chance of nearly a $60 per head loss. The probability of a profit is estimated at 57%. The financial analysis portion of the analysis utilized total returns to owner equity and risk. The optimistic equity return is $7,620, a 51% return on the owner's $15,000 equity commitment. Pessimistically, there is a one-in-six chance of a 33% equity loss. The expected return on equity is 15%, just over the 14% opportunity equity return of this producer.

RISK-RATED CASH FLOW BUDGETS

The risk-rated management strategy concepts developed by Ikerd and Anderson can be extended to multiple input price–output price–yield cost of production combinations. By introducing at least four types of variability in prices, yields, and costs, we may be generating too much mathematical complexity, so we might want to limit ourselves to just two variables or two sets of variables. There appears to be some justification for lumping together most yields into one optimistic-expected-pessimistic basket since a poor growing year for one crop will most likely imply a poor growing year for each crop. Similarly, many prices (especially crop prices) tend to be correlated during the growing season and from one year to the next. This correlation arises because government price policies and surpluses tend to move together and because weather is typically a macro variable, that is, it usually affects large regions and consequently large numbers of farmers similarly. A good growing season for one producer tends to be a good growing season for many producers, and many crops and the larger yields in aggregate push supply outward and prices downward.

"What If?" Budgets

Good cash flow planners will recognize the risk nature of their business and the variability of the major projections that they have made and prepare a series of risk-rated cash flow budgets or, stated in more common, simpler terms, a series of "What if?" cash flow budgets. Obviously with the very large number of input price, output price, yield, and cost of production estimates that were made in preparing the base and revised cash flow budgets, there are an extremely large number of possible "what if?" budgets that could be prepared. Consequently, we must find some way to provide a broad and meaningful but limited range of budgets to assess the risks that will be faced. Maintaining the probabilistic framework of the risk-rated management strategy concept is also important so that some degree of likelihood can be assessed.

One approach to limiting the number of budgets to a meaningful set is illustrated in Figure 10.3 (the best and worst outcomes may usually be ignored because their primary purpose was to avoid anchoring during the process of eliciting probabilities). Preparing the five budgets shown will provide a broad range of possible yet realistic outcomes. These budgets could, of course, be increased in number to include other price/yield combinations and/or other variables such as input prices and cost of production.

Picking the Plan or Assessing Risk

Assessing the effects on the cash flow budget of each of the risk-rated outcomes is the next necessary step. A number of financial measures should be calculated, including:[6]

1. $\dfrac{\text{Operating cash flow}[7]}{\text{Current liabilities}} = \dfrac{\text{OCF}}{\text{CL}}$

2. $\dfrac{\text{Operating cash flow}}{\text{Total liabilities}} = \dfrac{\text{OCF}}{\text{TL}}$

3. $\dfrac{\text{Operating cash flow}}{\text{Largest short-term credit balance}} = \dfrac{\text{OCF}}{\text{CB}}$

4. $\text{Current ratio} = \dfrac{\text{CA}}{\text{CL}}$

5. $\text{Rate of return on assets} = \dfrac{\text{Return to assets} \times 100}{\text{TA}}$

6. $\text{Rate of return on equity} = \dfrac{\text{Return to equity} \times 100}{\text{NW}}$

7. Ending cash balance

8. Ending operating loan balance

9. Net farm income

10. Ending net worth

If the likelihood of unfavorable outcomes is too high for our risk-bearing preferences, we must, unfortunately, return to Step 12 in the iterative planning process repeated in Chapter 9 and revise our marketing and production plans. Special attention should be paid to hedging with futures, options, and forward contracting, taking out crop insurance, and eliminating the most risky commodities from our whole-farm plan. Step 11 tells us to return to Step 2 and essentially begin again. The cash flow planning process may be necessary and may provide a great deal of information, but it is not necessarily easy.

Risk Management Rules of Thumb

Before we get totally discouraged by the great amount of work required in the cash flow planning process, be reminded that losses or failures *on paper* before production actually begins are highly preferable to *actual* losses. It is better to be aware of the chance of failure while there is still time to change your mind.

Anderson and Ikerd[8] close their risk management discussion with seven risk management rules of thumb that bear attention.

1. There are no certain profits.
2. Higher profits tend to be associated with higher risks of loss.
3. There is no one "best" strategy.
4. Good strategies improve the odds or chances for success.
5. Decisions should be evaluated with respect to objectives—not hindsight.
6. Good decisions sometimes have bad outcomes.
7. Good strategies, over time, will produce good results.

RISK-RATED CASH FLOW BUDGETS: SNAKE BELLY FARMS

Joe and Mary Farmer jumped into the assessment of the risks they were assuming with their final cash flow budget as prepared in Chapter 9 by estimating optimistic and

		PRICES		
		PESSIMISTIC 1/6	EXPECTED 2/3	OPTIMISTIC 1/6
Y I E L D S	1/6 (P)	Low Yield, Low Price (1/36)		Low Yield, High Price (1/36)
	2/3 (E)		EXPECTED YIELD, EXPECTED PRICE (4/9)	
	1/6 (0)	High Yield, Low Price (1/36)		High Yield, High Price (1/36)

Figure 10.3. Five-budget approach to risk-rated cash flow budget.

pessimistic input prices, output prices, and yields in accordance with the Anderson-Ikerd risk-rated management strategy concept. They estimated optimistic and pessimistic levels of each major variable as having a one-sixth chance that a more extreme level would occur. Their estimates are shown in Figure 10.4. A series of five output price/yield risk-rated cash flow budgets were prepared. The result of the five budgets in terms of the ten major financial measures noted earlier are presented in Figure 10.5.

Although undesirable, the low yield or low price outcome budgets were judged acceptable given the likelihood that each would occur.

SUMMARY

Assessment of the risks that must be borne to receive profits is a serious and difficult but necessary part of the forward planning process. The major sources of risk in agriculture are:

1. Production and yield risk
2. Market and price risk
3. Business and financial risk
4. Technology and obsolescence risk
5. Casualty loss risk
6. Social and legal risk
7. Human risk

While all seven are important and bear significant consideration when formulating long-term goals and plans, it is the first two or three that are most evident when developing short-run or annual operational plans.

	--------------Possible Outcomes------------		
Commodity	Optimistic	Expected	Pessimistic
	- - - - - - - - Input Prices - - - - - - - -		
Diesel fuel	$0.95/gal.	$1.00/gal.	$1.15/gal.
	- - - - - - - - Output Prices - - - - - - - -		
Corn	$2.25/bu.	$2.05/bu.	$1.90/bu.
Wheat	2.50/bu.	2.25/bu.	1.95/bu.
Cotton	0.80/lb.	0.60/lb.	0.45/lb
Alfalfa	100/ton	85/ton	65/ton
Stockers	68/cwt.	62/cwt.	60/cwt.
	- - - - - - - - - Yield - - - - - - - - - -		
Corn	200 bu.	170 bu.	140 bu.
Wheat	80 bu.	70 bu.	45 bu.
Cotton	900 lbs.	750 lbs.	550 lbs.
Alfalfa I	8 tons	7 tons	5 tons
Stockers	780 lbs.	740 lbs.	700 lbs.

Figure 10.4. Risk-rated optimistic, expected, and pessimistic outcomes, Snake Belly Farms.

	PRICES	
PESSIMISTIC	EXPECTED	OPTIMISTIC
1/6	2/3	1/6

1. OCF ÷ CL = -0.489		1. OCF ÷ CL = 0.319
2. OCF ÷ TL = -0.136		2. OCF ÷ TL = 0.061
3. OCF ÷ CB = -181.875		3. OCF ÷ CB = 73.176
1/6 4. CA ÷ CL = 0.559		4. CA ÷ CL = 1.152
(P) 5. RA% = -6.27%		5. RA% = 8.00%
6. RE% = -303.89%		6. RE% = 2.10%
7. ECB = 1,000		7. ECB = 1,000
8. EOLB = 155,583		8. EOLB = 14,827
9. NFI = -177,615		9. NFI = 21,448
10. NW = -16,604		10. NW = 182,459

Y

1. OCF ÷ CL = 0.292   OCF = Operating Cash Flow
2. OCF ÷ TL = 0.053   CL = Current Liabilities

I

3. OCF ÷ CB = 4.696   TL = Total Liabilities
4. CA ÷ CL = 1.201   CB = Largest Short-Term

E (1/6)

5. RA% = 8.64%   Credit Balance
6. RE% = 8.07%   CA = Current Assets

L (E)

7. ECB = 13,300   RA% = Rate of Return on
8. EOLB = 0   Assets

D

9. NFI = 29,727   RE% = Rate of Return on
10. NW = 145,344   Equity

S

ECB = Ending Operating
         Loan Balance
NFI = Net Farm Income
NW = Net Worth

1. OCF ÷ CL = -0.282		1. OCF ÷ CL = 1.066
2. OCF ÷ TL = -0.061		2. OCF ÷ TL = 0.194
3. OCF ÷ CB = -74.887		3. OCF ÷ CB = 1.616
1/6 4. CA ÷ CL = 0.784		4. CA ÷ CL = 1.887
(0) 5. RA% = 1.49%		5. RA% = 18.37%
6. RE% = -75.19%		6. RE% = 65.72%
7. ECB = 1,000		7. ECB = 141,312
8. EOLB = 50,181		8. EOLB = 0
9. NFI = -70,627		9. NFI = 176,735
10. NW = 90,384		10. NW = 337,746

Figure 10.5. Results of risk-rated cash flow budgets, Snake Belly Farms.

A particularly useful approach to assessing risk is the risk-rated management strategies concept that asks the planner to project optimistic, expected, and pessimistic estimates of all major risky variables. This concept can be extended to the development of several risk-rated cash flow budgets that can provide the planner and his or her financial adviser and lender with a picture of the effect or the consequences of adverse price or production conditions that might occur.

NOTES

1. A.G. Nelson, G.L. Casler, and O.L. Walker, *Making Farm Decisions in a Risky World: A Guidebook,* Extension Service, Oregon State University, Corvallis, July 1978, pp. 1-4.

2. K.B. Anderson and J.E. Ikerd, *Risk-Rated Management Strategies for Farm and Ranch Decisions,*

Circular E-841, Cooperative Extension Service, Oklahoma State University, Stillwater, Okla., 1984.

3. A more general, but more complicated, risk management methodology is Bayesian Decision Theory. See J.R. Anderson, J.L. Dillon, and B. Hardaker, *Agricultural Decision Analysis*, Iowa State University Press, Ames, Iowa, 1977.

4. An excellent set of learning games is provided in K.B. Anderson and J.E. Ikerd, *Risk: The Game for Survival and Profit*, Circular E-830, Cooperative Extension Service, Oklahoma State University, Stillwater, Okla., November 1983.

5. K.B. Anderson and J.E. Ikerd, *Risk-Rated Management Strategies for Farm and Ranch Decision*, p. V-9.

6. Parts of this list of financial measures were drawn from C.J. Casey and N.J. Bartczak, "Cash Flow—It's Not the Bottom Line," *Howard Business Review*, vol. 84, no. 4, pp. 60-66.

7. Operating cash flow is defined as net income + depreciation − withdrawals − increase in accounts receivable − increase in inventories + increase in accounts payable + increase in accrued liabilities.

8. K.B. Anderson and J.E. Ikerd, *Risk-Rated Management Strategies for Farm and Ranch Decisions*, p. vii-1.

RECOMMENDED READINGS

Kim B. Anderson and John E. Ikerd, *Risk-Rated Management Strategies for Farm and Ranch Decisions*, Extension Circular E-841, Cooperative Extension Service, Oklahoma State University, Stillwater, Okla., 1984.

John E. Ikerd and Kim B. Anderson, *Risk-Rated Management Strategies for Farmers and Ranchers*, OSU Extension Facts No. 159, Cooperative Extension Service, Oklahoma State University, Stillwater, Okla., December 1983.

Kim B. Anderson and John E. Ikerd, *Risk: The Game for Survival and Profit*, Extension Circular E-830, Cooperative Extension Service, Oklahoma State University, Stillwater, Okla., November 1983.

Jock R. Anderson, John L. Dillon, and Brian Hardaker, *Agricultural Decision Analysis*, Iowa State University Press, Ames, Iowa, 1977.

A. Gene Nelson, George L. Casler, and Odell L. Walker, *Making Farm Decisions in a Risky World: A Guidebook*, Extension Service, Oregon State University, Corvallis, Oreg., July 1978.

CHAPTER 11

Cash Flow Monitoring

Once a comprehensive cash flow plan has been completed and any revisions made, you can ease back in your chair, relax, and contemplate the extensive exercise you have just completed. You now have a much clearer picture of the status of your operation and a plan of procedure for the next operating period. With your goals clearly defined and this year's marketing, purchasing, and production strategies identified, you are now ready to move forward and put your plan into action.

Referring again to the analogy of the travelers in Chapters 1 and 2, the preliminary preparations have been made and it is now time to get in the car and go. However, a wise traveler will not forget his map. Though the destination has been clearly identified and the desired routes established, occasional reference to the map will tell the travelers where they are in relation to where they want to be. Are we on course? Are we driving enough miles each day to reach our destination at the scheduled time? Have we encountered obstacles that are not clearly identified on our map? What is the best way to get around this detour and stay on or near our marked course?

Just as these things are of concern to travelers, so should they be to cash flow planners. To forge ahead without reference to the cash flow plan or map is to say that I can get there from memory and there really won't be any problems along the way. Good luck!

PURPOSE OF CASH FLOW MONITORING

When a projected cash flow plan is completed at the beginning of the year, many of the factors involved are based on our best estimates rather than fact. The information we have to plan with and our ability to accurately project the future is at best imperfect and often times is severely limited. This may be more true for farmers and ranchers because of the numerous forces over which they have no control that can influence the outcome of their plans.

This does not mean, however, that the planning process and time involved is a wasted effort. Only if a plan is made and then never used or referred to is it a waste of time. Even so, many plans or ideas that are conceived are later rejected but may be the starting point or springboard to a better way of reaching goals. By going through the planning process we become aware of the problem areas, limitations, and factors most likely to change. As we become aware of these things we can explore what-if situations and formulate in the process alternative actions to be taken if the preferred course or outcomes do not occur.

Adopting the attitude that "the only constant is that things will change" should help us see the need for monitoring actual performance and comparing that with our projections. Generally, if a person is willing to go to the time and trouble to do an extensive cash flow plan, he or she will want to know how close his or her other projections are to actual performance. Without monitoring our position, it is easy to drift off course away from the stated objectives. Monitoring the cash flow of the business on at least a monthly basis will allow adjustments and corrections to be made when problems occur—and problems will occur. By ignoring this inevitability, farmers and ranchers would never reach their long-term goals but would wander aimlessly and "tread water" until they had exhausted their resources and could no longer survive.

THE CONTROL FUNCTION

The monitoring process for a business can be compared to the steering wheel of the car our travelers have been using. When traveling, we encounter curves, rough spots, oncoming traffic, and debris on the road that must be avoided. The steering wheel is the mechanism we use to control or guide the vehicle through or around these obstacles. The control function for a business is just as important for the well-being and survival of the firm as the steering wheel is to the traveler. This is our means of assuring that we will reach our intended destination or goals that have previously been defined.

Even the most experienced planners will not be able to make cash flow projections that will not require adjustment from time to time. Many lenders and others will say that farmers and ranchers can predict fairly accurately what their total costs will be for seed, fertilizer, and other variable costs. They know ahead of time how many acres will be planted, what the application rates will be, and generally what the prices of the inputs will be. The most difficult items to accurately project are crop and livestock receipts (due to price and yield variability), repairs, capital expenditures, and purchased livestock and feed costs. However, some of the variability in these items may be reduced or eliminated by using marketing tools as described in Chapter 5.

Though we may accurately project what total receipts and expenditures will be, the timing of receipts and payments will tend to vary somewhat from our projections. A good monitoring and control process will help ensure that the need for corrections and adjustments are detected and appropriate action then taken. It is easier to make small adjustments when needed than it is to make major adjustments that may be required later if problems are ignored or unseen.

To help illustrate this point and demonstrate the value of the control process, let us consider two examples—one requiring minor adjustments and one requiring major changes in the cash flow projections.

In the first example, Joe Farmer had planned to buy a new pickup truck that was budgeted in the cash flow for October. Joe had planned to give his 1975 model pickup to his hired hand to replace the old '68 Ford that the employee had been driving. However, the old Ford threw a rod in June and needed extensive work and expense to get it running.

Rather than spending money on the old truck, Joe decided to purchase the new pickup now and give his used one to the hired hand. After reviewing his cash flow plan, Joe determined that he had three options. He could sell the hay which he had planned to sell in October, he could reduce the amount of the repayment on the operating note in June, or he could borrow the funds now that were scheduled for October to buy the pickup. Joe decided to go ahead and borrow the funds for the pickup and hold the hay for a higher price in October.

Although the total cash flow projection for the year was unaffected, the timing of the pickup purchase created the need to make adjustments. Because Joe Farmer had a projected cash flow plan that he monitored, he was able to quickly make the necessary adjustment. He was also aware of how the change affected his business for the remaining part of the year.

In the second example things have taken a much more serious turn. Two weeks before harvesting his cotton, Joe's area was hit with a severe hail storm. When it was over Joe assessed his crop and determined overall damage of about 75 percent on his cotton. Earlier in the year Joe had considered buying crop insurance but had decided against it, because his area had been relatively free from hail storms and disasters for several years.

Now Joe was faced with a serious problem. The proceeds from his cotton crop were to be used to make a large portion of the payment on his land notes due in December. Joe had some hay in inventory that he planned to sell in December to help cover expenses the next spring. He also had some credit reserve available but wasn't exactly sure how his lender would want to handle the problem.

This situation calls for a reevaluation of the total operation. Joe needs a current balance sheet to accurately determine what his financial condition is at the present time. He must also begin to make a cash flow projection for the next year to determine what the potential is for making up the deficit caused by the hail storm. Many questions come to mind. Should he sell the hay now and see how he gets by next spring? Should he borrow the money to make the land payment? Would this note need to be scheduled over several years? How will this affect his cash flow next year?

As demonstrated by the two examples, cash flow monitoring should occur on a regular basis. Financial control of a business is in large part lost without regular monitoring and assessment of the impact of unforeseen events. The monitoring process keeps goals in mind and helps maintain the flexibility required to make needed adjustments.

MECHANICS

The monitoring process is relatively simple in contrast to making the initial cash flow projections. The control mechanism for monitoring cash flow projections is a monthly worksheet similar to the one shown in Figure 11.1. The worksheet used should follow the same format as the cash flow statement and should allow for comparison of actual versus projected cash flow.

The numbers in the amount budgeted column of the cash flow monitoring worksheet come directly from the projected cash flow for the appropriate month. This involves a minimal amount of time in transferring the budget figures to the monitoring worksheet. The numbers in the actual results column come directly from the monthly records summary. This again is a simple transfer of numbers from one page to another. However, depending on the type of records system you use, some additional time and work may be involved in totaling and transferring your monthly transactions to the worksheet.

There are several advantages to using a cash flow monitoring worksheet. The most obvious and most important of these is the comparison of actual results versus projections. As you calculate the variance (budgeted minus actual) for each month and the year-to-date totals, you will begin to see problems arise and the effects of your actions in response.

In addition to the above comparison of actual versus budgeted amounts, a cash flow monitoring worksheet can prove very useful in finding errors in record books. This is especially true if a single-entry records system, which lacks many of the checks and balances of double-entry systems, is used. If the cash flow monitoring worksheet is done consistently at the end of each month, any errors can be corrected at that time. This is much easier than trying to remember what happened six months ago when catch-up time rolls around.

The cash sources = cash uses equation must balance every month. If it does not, this is an immediate indication of errors. The errors could be math errors or incorrect totals on the monitoring worksheet. After double-checking the numbers and math to see that everything was transferred and totaled correctly on the worksheet, if cash sources still do not equal cash uses, errors in the records system are indicated. Such errors could be omitted transactions, double entries, or transposition and math errors. Completing the monitoring worksheet every month and verifying that cash sources equals cash uses will help assure completeness and accuracy in your records.

A completed cash flow monitoring worksheet may also provide much of the information needed for tax planning during the year. In addition it makes a quick reference for providing needed information to the operator, landlord, and/or lenders.

It is usually easiest to follow the worksheet format to complete the actual results column of the monitoring worksheet. The type of records system and monthly summary you have will help determine the easiest way to proceed. The beginning cash balance is easy—it is the ending cash balance from the previous month.

After entering the beginning cash balance amount, proceed by entering the operating receipts, capital receipts, and nonfarm receipts received that month. Then total these amounts to get total cash available and record this on the appropriate line. Next, enter the operating expenses, livestock and feed purchases, capital, and other expenditures, including intermediate and long-term principal and interest payments. These numbers are then totaled to arrive at total cash required on the worksheet.

The last section of the worksheet should be completed by filling in transfers into and out of savings, money borrowed, and operating loan principal and interest payments. Enter the ending cash balance from your check register or bank statement

MONITORING WORKSHEET

Item from Cash Flow Budget	Month of: Amount Budgeted	Actual Results	Variance (+ or −)	Year-to-Date Totals Amount Budgeted	Actual Results	Variance (+ or −)
1 Beginning cash balance						
Operating Receipts:						
2 Feeder livestock and poultry						
3 Crops and feed						
4 Livestock and poultry products						
5 Custom work; cash patronage dividends						
6 Government payments (cash)						
7 Hedging account withdrawals						
8						
Capital Receipts:						
9 Breeding livestock						
10 Machinery; equipment; real estate						
11						
Non-Farm Income						
12 Off-farm wages						
13 Interest and dividends						
14-15 Other businesses & investments						
16 TOTAL CASH AVAILABLE (add lines 1 thru 15)						
Operating Expenses:						
17 Chemicals						
18 Custom machine hire						
19 Fertilizer and lime						
20 Gas, fuel, oil						
21 Insurance (property, liability, crop)						
22 Labor hired						
23 Livestock expenses (breeding, vet, etc.)						
24 Marketing and transportation						
25 Rents and leases						
26 Repairs—machinery and equipment						
27 —buildings and improvements						
28 Seed						
29 Storage and custom drying						
30 Supplies						
31 Taxes (real estate and personal property)						
32 Utilities (farm share)						
33 Auto (farm share)						
34-35						
36 Total Cash Operating Expenses						

Figure 11.1. Cash flow monitoring worksheet, AFRA Schedule C-9.

Source: A.W. Oltmans, D.A. Klinefelter, and T.L. Frey, *Agricultural Financial Reporting and Analysis,* Century Communications, Inc, Niles, Ill., 1992.

Item from Cash Flow Budget	Month of:			Year-to-Date Totals		
	Amount Budgeted	Actual Results	Variance (+ or −)	Amount Budgeted	Actual Results	Variance (+ or −)
Livestock and Feed Purchases:						
37 Feeder Livestock						
38 Breeding Livestock						
39 Feed purchased						
Capital Expenditures:						
40 Machinery and equipment						
41 Buildings and improvements						
42						
Other Expenditures:						
43 Hedging account deposits						
44 Family living withdrawals						
45 Other businesses & investments						
46 Income tax and social security						
47						
48 Term debt loan payments – principal						
49 – interest						
50 **TOTAL CASH REQUIRED** (add lines 36 thru 49)						
51 **CASH AVAILABLE LESS CASH REQUIRED** (line 16 minus 50)						
52 Inflows from savings						
53 Cash position before borrowing						
Money to be borrowed						
54 – operating loans						
– term debt						
55 Operating loan payments – principal						
– interest						
57 Outflows to savings						
58 Ending cash balance						
Loan Balances: (at end of period)						
59 Current year's operating loans						
60 Previous year's operating loans						
61 Term debt loans						
62 Total Loans						

Figure 11.1. (*continued*).

This form is copyrighted. It is a violation of the U.S. Copyright Law to reproduce it in any manner. To order forms write or call Century Communications Inc., 6201 Howard St., Niles, IL 60714; 708/647-1200 or Doane Agricultural Services Co., 11701 Borman Dr., St. Louis, MO 63146; 314/569-2700.

depending on how your records are kept. If you record your checks in your record book as they are written, the ending cash balance on the worksheet should agree with the ending cash balance in your check register. If you record your checks as they are returned with your bank statement, the ending bank statement balance and ending cash balance on the worksheet should agree.

A consistency check should be done now to ensure that total cash sources equals total cash uses. If the two are not equal, you must find any errors and correct them. When the worksheet is correct and sources equals uses, you are ready to calculate the monthly variance and year-to-date totals. The year-to-date total sources and uses should be equal just as each individual month should be. Any differences between year-to-date sources and uses would indicate math errors in totaling the year-to-date figures (if all months have been completed correctly).

If significant changes have occurred in the actual cash flow from what was projected, the cash flow for the remainder of the year may need to be reprojected. This is especially true if changes have occurred that would affect debt repayment capacity or loan repayment schedules.

VARIANCE ANALYSIS

Once the mechanical process of computing monthly and year-to-date budget variances has been completed, the real work can begin. The challenge in analyzing variances is two-fold: 1) to determine what caused the variance; and 2) to determine whether it will occur again. In the management accounting literature, this process of variance analysis falls under a general heading of responsibility accounting and can be broken into two general pieces—flexible budgets and basic variance analysis measures.[1]

The concept of a flexible budget is difficult to properly adapt to an agricultural production process, especially because most agricultural commodities complete only one production cycle during the year. In general, the flexible budget adjusts sales revenues and variable costs to the amount of units actually produced rather than forcing the budgeted inflow and outflow amounts to remain static at the physical unit level originally budgeted or forecast. The difficulty in application comes from the following three different directions:

1. With only one production cycle during the year, monthly budgets cannot be flexed until after the end of the cycle.
2. Few agricultural inputs are truly variable on a per bushel or per pound produced basis (i.e., if one quart of insecticide per acre increases corn yields by 20 bushels per acre, will one gallon increase yields by 80 or even any at all?).
3. A great deal of the variability in the output amount is caused by external forces, most notably weather, rather than by changed output goals, changed input amounts, or different levels of sales or orders.

Regardless of these difficulties, budgeted amounts should be adjusted for known changes before comparing the budgeted amounts to actual. The impact on future amounts within the overall cash flow budget must be considered, providing even more reason to keep an alive, ready-to-be-revised cash flow plan.

A variety of variance measures have been designed for and are used in analysis of accounting information in non-agricultural businesses. Most were developed for manufacturing firms, but some help analyze retail or wholesale firms as well. Regardless of the type of industry for which each variance measure was developed, each has a common premise—actual results seldom, if ever, come out the same as budgeted or planned. The goal of each variance measure is to find the reason why actual results differ from planned levels. Various questions are posed in order to get at this reason. Assume for a moment that the expenditures for a basic input, such as fuel, were larger than expected. Were prices higher than expected, leading to the high expenditures? Or was the use of fuel-consuming machinery greater than expected? If dealing with a cash budget, we might also need to ask if lower prices led the manager to buy more and store it for use in a later period.

A typical variance analysis is applied to the cost of direct materials used in the production process and includes analysis of both price and quantity differences. The materials price variance is defined as the difference between the actual materials cost and the budgeted cost of actual materials inputs. The materials quantity variance is defined as the difference between the budgeted cost of actual materials inputs and the budgeted cost of materials (from the flexible budget). Added together, these two variances show the total difference between actual expenditures and budgeted expenditures. Separated, they attempt to focus managerial attention on the cause of the difference to determine whether price or quantity is the culprit.

Consider the example shown in Figure 11.2 for diesel fuel. The manager budgeted $3,200 for fuel purchases during the month, expecting to use 3,200 gallons at $1.00 per gallon. However, expenditures actually amounted to $3,378 (3,157 gallons at $1.07 per gallon). The materials price variance, as defined previously, relates actual cost to expected cost of actual use; it abstracts from quantity differences to focus on differences caused solely by price changes.

$$\text{Materials price variance} = AQ(AP - BP)$$
$$= 3{,}157 \,(1.07 - 1.00)$$
$$= \$221 \text{ U}$$

Because actual price was higher than budgeted price by 7 cents, $221 more than expected was spent for 3,157 gallons, leading to the $221 unfavorable (U) variance.

The materials quantity variance relates expected cost of actual use to budgeted

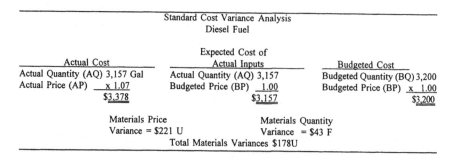

Figure 11.2. Example standard cost variance analysis.

cost; it abstracts from price differences to focus on quantity differences.

$$\begin{aligned}
\text{Materials quantity variance} &= BP(AQ - BQ) \\
&= 1.00\ (3,157 - 3,200) \\
&= 1.00\ (-43) \\
&= \$43\ F
\end{aligned}$$

Because actual use was less than budgeted use by 43 gallons, $43 less was spent than expected at $1.00 per gallon, leading to a $43 favorable (F) variance.

Combining the price and quantity variances, a $221 unfavorable and a $43 favorable variance, implies an overall unfavorable materials variance of $178. Interpretation of the overall variance is clarified by noting that the fuel use was lower than expected, implying good control, and by noting the negative impact of the price increase, indicating the need to control purchases, find alternate suppliers, or find further ways to conserve.

Many other variances, including similar variances for labor and overhead items as well as other materials, should be computed and analyzed.

MONITORING WORKSHEET: SNAKE BELLY FARMS

After completing and summarizing his farm records each month, Joe Farmer was able to easily complete the cash flow monitoring worksheet shown in Figure 11.3. The actual results column was completed directly from his monthly records summary, and the amount budgeted column was completed from his projected cash flow plan (Figure 3.3).

SUMMARY

A comprehensive cash flow plan will outline the specific details of a short-run financial plan by which intermediate and long-term goals will be reached. The monitoring and control process of the business will ensure that problems are detected and attended to before they become serious or fatal. It will also enable the operator to adjust and maintain the needed flexibility to manage the business and control inflows and outflows.

Because of uncertainties and imperfect information available when projections are made, proper financial control is vital to the success of any business. Proper control is provided through good business records, cash flow monitoring, and business analysis. The monitoring process will enable us to stay "on course" toward our intended goal or destination. Good records will tell us where we have been and how we got where we are. Good business analysis will tell us whether or not we are achieving our goals and what changes should be made for the long-run well-being of the business.

MONITORING WORKSHEET
Snake Belly Farms
Projected for 19X2 vs. Actual

PAGE 1

	MONTH OF MARCH			YEAR TO DATE		
ITEM FROM CASH FLOW	AMOUNT BUDGETED	ACTUAL RESULTS	VARIANCE	AMOUNT BUDGETED	ACTUAL RESULTS	VARIANCE
BEG. CASH BALANCE	1,000	1,405	405	1,000	1,405	405
OPERATING RECEIPTS:						
CROPS AND FEED	733	825	92	7,618	7,475	-143
LIVESTOCK & POULTRY	0	0	0	0	0	0
PRODUCTS (LIVESTOCK)	0	0	0	0	0	0
CUSTOM WORK	0	0	0	0	0	0
GOVERNMENT PAYMENTS	0	0	0	0	0	0
HEDGING ACCOUNT W/D	0	0	0	0	0	0
CAPITAL RECEIPTS:						
BREEDING STOCK	0	0	0	0	0	0
MACHINERY & EQUIPMENT	0	0	0	0	0	0
NON-FARM INCOME:						
OFF-FARM WAGES	1,200	1,200	0	3,600	3,600	0
INTEREST & DIVIDENDS	0	0	0	0	0	0
TOTAL CASH AVAILABLE	2,933	3,430	497	12,218	12,480	262
OPERATING EXPENSES:						
LABOR HIRED	850	850	0	2,550	2,550	0
REPAIRS-MACHINERY & EQUIPMENT	760	240	-520	2,280	2,370	90
REPAIRS-BUILDINGS/IMPROVEMENTS	75	0	-75	225	120	-105
RENT & LEASES	0	0	0	0	0	0
SEED	0	0	0	0	0	0
FERTILIZER & LIME	0	0	0	0	0	0
CHEMICALS	0	0	0	0	0	0
CUSTOM MACHINE HIRE	0	0	0	0	0	0
SUPPLIES	125	110	-15	375	420	45
LIVESTOCK EXPENSE	600	625	25	1,980	1,870	-110
GAS, FUEL, OIL	0	0	0	0	0	0
STORAGE/CUSTOM DRY	0	0	0	0	0	0
TAXES (REAL ESTATE, PP)	0	0	0	2,112	2,165	53
INSURANCE (PROP, LIAB)	1,000	1,000	0	1,800	1,800	0
UTILITIES (ELECTRIC/GAS)	170	165	-5	510	505	-5
MARKET/TRANSPORT EXPENSE	0	0	0	0	0	0
AUTO (FARM SHARE)	350	320	-30	1,050	1,005	-45
ACCOUNTS PAYABLE	0	0	0	7,000	7,000	0
EDUCATION, TRAINING, MISC	1,233	1,250	17	1,433	1,450	17
TOTAL CASH OPERATING EXPENSES	5,163	4,560	-603	32,225	32,615	390

Figure 11.3. Cash flow monitoring worksheet, Snake Belly Farms.

MONITORING WORKSHEET
Snake Belly Farms
Projected 19X2 vs. Actual   PAGE 2

ITEM FROM CASH FLOW	MONTH OF MARCH			YEAR TO DATE		
	AMOUNT BUDGETED	ACTUAL RESULTS	VARIANCE	AMOUNT BUDGETED	ACTUAL RESULTS	VARIANCE
STOCK & FEED PURCHASES:						
FEEDER LIVESTOCK	0	0	0	0	0	0
BREEDING LIVESTOCK	0	0	0	0	0	0
FEED PURCHASED	0	0	0	1,458	1,450	-8
CAPITAL EXPENDITURES:						
MACHINERY & EQUIPMENT	2,000	2,000	0	2,000	2,000	0
BUILDINGS & IMPROVEMENTS	0	0	0	0	0	0
OTHER EXPENDITURES:						
HEDGING ACCT DEPOSIT	0	0	0	0	0	0
GROSS FAMILY LIVING W/D	1,500	1,500	0	4,500	4,500	0
NON-FARM BUS/INVESTMENT	0	0	0	0	0	0
INCOME TAX & SOCIAL SECURITY	1,000	850	-150	1,000	850	-150
LOAN PAYMENTS - PRINCIPAL	0	0	0	0	0	0
LOAN PAYMENTS - INTEREST	0	0	0	0	0	0
TOTAL CASH REQUIRED	9,663	8,910	-753	41,183	41,415	232
CASH AVAILABLE - CASH REQUIRED	-6,730	-5,480	1,250	-28,965	-28,935	30
INFLOWS FROM SAVINGS	0	0	0	0	0	0
CASH POSITION BEFORE BORROWING	-6,730	-5,480	1,250	-28,965	-28,935	30
MONEY TO BE BORROWED						
- OPERATING LOANS	7,730	6,500	-1,230	10,055	10,000	-55
- INTEREST & L/T LOANS	0	0	0	0	0	0
OPERATING LOAN PAYMENT - PRINCIPAL	0	0	0	0	0	0
- INTEREST	0	0	0	0	0	0
OUTFLOWS TO SAVINGS	0	0	0	0	0	0
ENDING CASH BALANCE	1,000	1,020	20	1,000	1,020	20

Figure 11.3. (continued).

NOTE

1. Students are urged to take a course in managerial accounting to round out their analytical business education. Other readers should consult a managerial textbook for a more thorough discussion of both flexible budgeting and variance analysis. A good example is W.J. Morse, J.R. Davis, and A.L. Hartgraves, *Management Accounting*, 3rd ed., Addison-Wesley Publishing Company, Reading, Mass., 1991, Chapters 7 and 8.

RECOMMENDED READINGS

A. Gene Nelson and Thomas L. Frey, *You and Your Cash Flow*, Century Communications, Inc., Skokie, Ill., 1983.

Wayne J. Morse, James R. Davis, and Al L. Hartgraves, *Management Accounting*, Addison-Wesley Publishing Company, Reading, Mass., 3rd. ed., 1991.

PART IV

Limitations and Other Concerns

The risk in writing, teaching, reading, and following a discourse on the preparation of cash flow budgets for a farm or ranch is the tendency to develop a belief that the cash flow budget is the most important, and in the extreme, the only important financial planning device. Throughout this text we have stressed that cash flow planning is a critical element in the annual operational planning process, but we have attempted to condition our enthusiasm for development of cash flow budgets through repeated reminders that the cash flow budget is *not* all things to all people.

The cash flow budget does not address the issue of profitability, a poorly done cash flow budget is probably more dangerous than none at all, and the short-run operational planning process must be modified to analyze long-run decisions. Each of these problem areas is discussed in the last chapter of the book. The intent of this chapter is not to diminish the value of the cash flow planning process, but rather to place it in the context of other issues and analytical methods.

RECOMMENDED READINGS

Cornelius J. Casey and Norman J. Bartczak, "Cash Flow—It's Not the Bottom Line," *Harvard Business Review*, 84:4, July-August 1984, pp. 60-66.

David A. Lins, Daniel J. Seger, and Bruce Burk, "Cash-Income Fallacies," *Agri-Finance*, February 1987, pp. 64-66.

Michael D. Boehlje, "Performance-Based Landing," *Agri-Finance*, February 1987, pp. 68-69.

Jock R. Anderson, John L. Dillon, and Brian Hardaker, *Agricultural Decision Analysis: Using Discounted Cash Flows*, 4th ed., Cornell University, Ithaca, N.Y., 1988.

John B. Penson, Jr., Danny A. Klinefelter, and David A. Lins, *Farm Investment and Financial Analysis*, Prentice-Hall, Inc., Englewood Cliffs, N.J., 1982.

Farm Journal, "Cash Flow Not a Cure-All," *Farm Journal*, March 1985, p. BEEF-23.

CHAPTER 12

Cash Flow Deficiencies, Problems, and Long-Run Projections

The cash flow budget is not the storehouse of all important financial information. Rather, it is only an excellent device for leading us to the development of a comprehensive set of financial information. In this chapter we will try to reemphasize that cash flow budgets are only one component of the planning process, reemphasize problems in projecting cash flows, and briefly discuss the nature of longer-run cash flow planning procedures.

PROFITABILITY VERSUS LIQUIDITY

Simply stated, the cash flow budget tells us *nothing* about the profitability of the business; it merely describes the liquidity of the business. While it is absolutely critical in the short run for every business to generate sufficient positive cash flow to meet all of its financial commitments, to pay its operating expenses, and to provide sufficient excess to satisfy the needs of the owner and his family, it is equally as critical, if not more so, that the business generate sufficient profits in the long run to grow or to sustain itself, to replace worn out or obsolete capital, and to meet the owner's long-run needs and goals. The cash flow plan, as described early in this book, is analogous to a road map when taking a long trip. The map plots the ways and means for getting from one point to another. The cash flow plan summarizes operational and financial plans and shows how the business can move from January 1 to December 31, all the while living within the rules of the game: cash balances must never be negative, and the physical and credit limitations imposed upon the business must not be exceeded.

Final judgment of the plan, however, must be assessed using accrual net income developed on the income statement and the related effect on net worth generated on the balance sheet. Only the accrual net income calculation provides a guide to the profitability of the business.

We noted earlier that there are several avenues for making a cash flow plan work. These avenues include the following:

1. Subsidizing the business with off-farm or off-ranch income
2. Selling capital assets
3. Borrowing capital or restructuring debt
4. Selling crop and livestock inventories
5. Drawing on savings accounts
6. Relying upon other sources of outside cash, such as gifts or inheritances

Each of these avenues might be used to meet a deficit in cash available less cash required, and they may be necessary and used without guilt over short periods of time, especially when the business is expanding or experiencing short-term financial difficulties. But continued reliance upon these cash sources will hide the underlying problem that the business is simply not profitable. Only the accrual net income figure can sort out the financial effects of continued reliance upon outside revenue sources. In short, it is possible to generate positive cash flow and still experience a protracted period of net losses and declining net worth. Accumulated net losses will eventually lead to financial distress and possibly to bankruptcy.

Consider the study conducted by Casey and Bartczak[1] concerning excessive reliance upon cash flow information to gauge the financial health of a business. Although dealing with nonfarm business, their findings are applicable to our situation. They noted that:

> A growing number of securities analysts, financial writers, and accounting policymakers contend that financial statements providing information of a company's cash flows yield a better measure of operating performance than do the company's income statement and balance sheet. According to recent surveys, corporate and government officials have accepted this view; they rated cash flow data the most important piece of information contained in published financial statements.
>
> Accordingly, securities analysts have come to view cash flow information as a more accurate yardstick for gauging debt and dividend-paying ability. Corporate executives have penetrated the veil of accounting profits, have found them sometimes misleading, and have turned to the "real thing," cash flow data, to evaluate their company's performance and that of competitors.

They studied 60 companies that filed bankruptcy petitions between 1971 and 1982 and compared their financial reports to those of 230 viable companies. They conclude that:

> The results suggest that other factors, such as a company's debt level, its access to the debt and equity markets, the salability of its capital assets, and its reservoir of liquid assets, may be better indicators of its survival prospects than cash flow data.
>
> Our finding that operating cash flow data do not accurately distinguish between healthy companies and dying ones raises a question about the presumed value of cash flow data for analyzing and forecasting a company's performance. Elevating cash flow, without testing its applicability, as the panacea for the problem of assessing performance is akin to the euphoria in the 1960s surrounding growth in earnings per share as supposedly the best indicator of company value. We hope that unbridled enthusiasm for cash flow data will not produce a repeat of the debacles that resulted from blindly following earnings per share growth.[2]

Similarly, in arguing that lenders should not overemphasize cash flow lending, Boehlje[3] suggests that, "A preferable strategy for today's conditions is performance-based lending—lending decisions based on collateral and risk-bearing ability, cash flow and repayment capacity, ability to generate earnings, as well as economic efficiency and business performance." He emphasizes that performance-based lending (a concept that could be directly extended to performance-based planning and analysis) should be based on the following four key components:

1. Risk and collateral assessment
2. An assessment of the income-generating capacity of the business
3. Business efficiency and productivity analysis
4. Assessing repayment capacity[4]

INABILITY TO FORECAST

A second major problem with developing and relying upon cash flow budgets is our inability to forecast the future with a significant degree of accuracy. Forecasting yields and prices is an extremely imprecise process, because we simply cannot forecast all of the weather, supply and demand, government, labor, political, and international factors that influence our actual price and output outcomes.

We can, of course, rely upon the forecasts of experts, but those experts are also unable to project all of the price and output influences. Anderson and Ikerd summarized the accuracy of major price forecasts; those results are shown in Figures 12.1 and 12.2.[5]

OUTLOOK SOURCE	CURRENT QUARTER	ONE QUARTER IN ADVANCE	TWO-QUARTERS IN ADVANCE	THREE QUARTERS IN ADVANCE
1979				
MAJOR PRIVATE FIRMS	---	10-13%	12-19%	13-25%
FUTURES MARKET	---	10%	18%	
1989 O.S.U. STUDY				
UNIVERSITY ECONO-MISTS	5-6%	11-13%	12-15%	11-13%
USDA	5%	11%	18%	---
O.S.U.	5%	11%	12%	11%
FUTURES O.S.U.	5%	11%	14%	17%

Figure 12.1. Accuracy of cattle outlook information: forecast errors of various organizations.

Source: K.B. Anderson and J.E. Ikerd, *Risk-Rated Management Strategies for Farm and Ranch Decisions,* Extension Circular E-841, Cooperative Extension Service, Oklahoma State University, Stillwater, Okla., 1984.

Perhaps the best, or even only, way to counteract or recognize our inability to forecast the future with perfect accuracy is to prepare a series of risk-rated cash flow budgets as discussed in Chapter 10 and to monitor and continually revise our cash flow budgets as more information becomes available as described in Chapter 11. We might also investigate more fully the Bayesian decision theory procedures that allow us to incorporate the type of accuracy or reliability of forecast data as shown in Figures 12.1 and 12.2.[6]

UNJUSTIFIED OPTIMISM

A third major problem with cash flow forecasts follows the difficulty in forecasting—it is our tendency to be overly optimistic. One study found that a group of borrowers overestimated cash receipts by 15 percent.[7] The tendency towards wishful thinking can be significantly curtailed through the ground-up development process described in the first 11 chapters of this book. Much of the unjustified optimism falls out from a one-page cash flow budget developed to keep the lender happy.

LONG-RUN CASH FLOW PLANNING

The preceding eleven chapters focused on the cash flow plan for a single year. The purpose of a one-year detailed cash flow plan is to project the cash operating needs of the business for the next year. The detail and timing in these projections is important because it will affect the borrowing needs and repayment ability for the next production cycle. However, at times the need arises to do some cash flow projections for longer periods. Such projections would be called for when considering the purchase of an additional farm or ranch, the purchase of major equipment, changing or adding enterprises, debt restructuring, or other major considerations. Normally, in these situations, we want to evaluate the overall effects on the business of the action being considered. Therefore we are not so concerned with the detailed day-to-day timing of cash inflows and outflows as we are in determining the net cash flow to the business after the changes have been made.

Commodity	1 Mo.	3 Mo.	6 Mo.	9 Mo.
Cattle:				
Slaughter	5%	11%	14%	17%
Feeders	7%	10%	16%	19%
Wheat:	10%	13%	15%	18%
Soybeans:	na.	15%	20%	25%
Cotton:	na.	15%	22%	23%
Corn:	na.	13%	16%	19%
Hogs:	na.	10%	17%	20%

Figure 12.2. Price forecast accuracy: forecast errors using futures prices.
Source: K.B. Anderson and J.E. Ikerd, *Risk-Rated Management Strategies for Farm and Ranch Decisions,* Extension Circular E-841, Cooperative Extension Service, Oklahoma State University, Stillwater, Okla., 1984.

For cash flow projections beyond one year, an annual or quarterly cash flow projection will provide sufficient information to make long-run decisions. An annual or quarterly cash flow would tell us, for example, if the debt payments under a restructuring plan could be serviced or if the cash requirements of a new or expanded enterprise can be met.

Long-Run Planning Methods

When planning for the short-run we are concerned with using actual facts where possible and accurate projections in determining crop and livestock prices, yields, and input prices. While this approach is necessary but difficult for short-run planning, it is impractical for long-run projections. When planning for a three- to five-year period we generally want to evaluate the changes that will occur in the business due to new investments or production practices. Therefore, because of the difficulty in predicting future events, prices, and yields, we will normally assume that "average" or "normal year" conditions will prevail for the planning period.

This can be done by using five-year average commodity prices, yields, and input prices. Known economic trends such as increasing or decreasing oil prices also should be considered in the planning process. Such trends will directly affect input prices for items such as fuel, fertilizer, and chemicals.

The planning process and format for long-run cash flow planning is the same as the short-run processes described in this book and used in the Snake Belly Farms example. After determining acceptable yields and prices, total farm operating receipts can be projected using planned crop rotation and livestock practices. Non-farm income and capital sales should also be estimated for each phase of the planning period. Farm operating expenses, capital expenditures, non-farm expenditures, and intermediate and long-term debt payments should be projected based on determined input prices and planned investments. Existing intermediate and long-term debt payments for the planning period are known and payments due on new loans can be easily determined.

The next step is to determine the operating loan requirements and repayment ability for each phase or period of the planning horizon. Only when this last step is completed will we be able to clearly see the overall effects that a new investment or other change would have on the total cash flow requirements of the business.

Discounted Cash Flows

Cash flow analysis of proposed business investments is useful and important but also has some limitations. One of the major shortcomings of cash flow projections in evaluating new investments is that the time value of money is ignored. Very often alternative investments being considered have different lives and require cash investments at different time periods. Also ignored in a cash flow projection are the effects of inflation, taxes, and risk.

In order to compare alternative investments with varying returns and cash requirements, it is useful to convert all cash flows to the same time period. This can be done by discounting future cash flows to the present. In this way, investments may be more accurately compared to alternatives.

It is beyond the scope of this textbook to provide a detailed discussion of discounting methods and their strengths and weaknesses. The reader is referred to other texts for an in-depth look at investment analysis.[8]

By combining cash flow analysis with other investment analysis techniques, we can get a clearer picture of the relative benefits of alternative investment choices. Many investments, such as a pecan orchard, will have a negative cash flow for the first few years. Based on cash flow analysis alone, this would not appear to be a wise investment. However, a net present value analysis of the investment may show the pecan orchard to meet or exceed management's criteria for accepting one investment over another.

NOTES

1. C.J. Casey and N.J. Bartczak, "Cash Flow—It's Not the Bottom Line," *Harvard Business Review*, vol. 84, no. 4, July-August 1984, p. 61.

2. Ibid., p. 65.

3. M.D. Boehlje, "Performance-Based Lending," *Agri-Finance*, Century Communications, Skokie, Ill., February 1987, p. 68.

4. A further discussion of business efficiency and productivity analysis can be found in J.D. Libbin and L.B. Catlett, *Farm and Ranch Financial Records*, Macmillan, New York, N.Y., 1987.

5. See K.B. Anderson and J.E. Ikerd, *Risk-Rated Management Strategies for Farm and Ranch Decisions*, Circular E-841, Cooperative Extension Service, Oklahoma State University, Stillwater, Okla., 1984.

6. See J.R. Anderson, J.L. Dillon, and B. Hardaker, *Agricultural Decision Analysis*, Iowa State University Press, Ames, Iowa, 1977.

7. See D.A. Klinefelter, *Restructuring the Farm Business*, B-1549, Texas Agricultural Extension Service, Texas A&M University, College Station, May 1989.

8. See G.L. Casler, B.L. Anderson, and R.D. Aplin, *Capital Investment Analysis: Using Discounted Cash Flow*, 4th ed., Cornell University, Ithaca, N.Y., 1988; and J.B. Penson, Jr., D.A. Klinefelter, and D.A. Lins, *Farm Investment and Financial Analysis*, Prentice-Hall, Englewood Cliffs, N.J., 1982.

RECOMMENDED READINGS

Cornelius J. Casey and Norman J. Bartczak, "Cash Flow—It's Not the Bottom Line," Harvard Business Review, 84:4, July-August 1984, pp. 60-66.

David A. Lins, Daniel J. Seger, and Bruce Burk, "Cash-Income Fallacies," *Agri-Finance*, February 1987, pp. 64-66.

Michael D. Boehlje, "Performance-Based Landing," *Agri-Finance*, February 1987, pp. 68-69.

Jock R. Anderson, John L. Dillon, and Brian Hardaker, *Agricultural Decision Analysis*, Iowa State University Press, Ames, Iowa, 1977.

George L. Casler,, Bruce L. Anderson, and Richard D. Aplin, *Capital Investment Analysis: Using Discounted Cash Flows*, 4th ed., Cornell University, Ithaca, N.Y., 1988.

John B. Penson, Jr., Danny A. Klinefelter, and David A. Lins, *Farm Investment and Financial Analysis*, Prentice-Hall, Inc., Englewood Cliffs, N.J., 1982.

Farm Journal, "Cash Flow Not a Cure-All," *Farm Journal*, March 1985, p. BEEF-23.

INDEX